DRYING

IN THE SAME SERIES

Oil Processing
Fruit and Vegetable Processing
Root Crop Processing
Fish Processing
Cereal Processing
Storage
Women's Roles in Technical Innovation

Food Cycle Technology Source Books

DRYING

INTERMEDIATE TECHNOLOGY PUBLICATIONS
in association with the
United Nations Development Fund for Women (UNIFEM) 1995

Intermediate Technology Publications Ltd,
103–105 Southampton Row, London WC1B 4HH, UK

A CIP record for this book is available from the British Library

ISBN 1 85339 308 8

Illustrations by Matthew Whitton, UK

Typeset by Dorwyn Ltd, Rowlands Castle. Printed in Great Britain by SRP, Exeter

Contents

Preface

This source book is one of a continuing UNIFEM series which aims to increase awareness of the range of technological options and sources of expertise, as well as indicating the complex nature of designing and successfully implementing technology development and dissemination programmes.

UNIFEM was established in 1976, and is an autonomous body associated since 1984 with the United Nations Development Programme. UNIFEM seeks to free women from underproductive tasks and augment the productivity of their work as a means of accelerating the development process. It does this through funding specific women's projects which yield direct benefits and through actions directed to ensure that all development policies, plans, programmes and projects take account of the needs of women producers.

In recognition of women's special roles in the production, processing, storage, preparation and marketing of food, UNIFEM initiated a Food Cycle Technology project in 1985 with the aim of promoting the widespread diffusion of tested technologies to increase the productivity of women's labour in this sector. While global in perspective, the initial phase of the project was implemented in Africa in view of the concern over food security in many countries of the region.

A careful evaluation of the Africa experience in the final phase of this five-year programme showed that there was a need for catalytic interventions which would lead to an enabling environment for women to have easier access to technologies. This would be an environment where women producers could obtain information on the available technologies, have the capacity to analyse such information and make technological choices on their own, as well as the capacity to acquire credit and training to enable their purchase and operation of the technology of their choice. This UNIFEM source book series aims to facilitate the building of such an environment.

Acknowledgements

This series of food cycle technology source books has been prepared at Intermediate Technology (IT) in the United Kingdom within the context of UNIFEM's Women and Food Cycle Technologies specialization. Translation and printing of the books is undertaken by the Italian Association for Women in Development (AIDOS).

During the preparation process the project staff contacted numerous project directors, rural development agencies, technology centres, women's organizations, equipment manufacturers and researchers in all parts of the world.

The authors wish to thank the many agencies and individuals who have contributed to the preparation of this source book. In addition to those listed in the Contacts section, special thanks are owed to David Trim of the Natural Resources Institute, UK, Helen Appleton and Emma Crewe of IT, UK, Bertha Msora of Ranche House College, Zimbabwe, Abdullah Al Mahmud of the Mennonite Central Committee, Bangladesh, and Jim McDowell, Independent Consultant, UK, for their invaluable contributions.

The preparation of the first five source books was funded by UNIFEM with a cost-sharing contribution from the Government of Italy and the Government of the Netherlands. The Government of Italy provided the funds for continuation of the series, as well as the translation of the books into French and Portuguese and printing of the first editions.

Peggy Oti-Boateng
UNIFEM consultant
University of Science and Technology
Kumasi, Ghana

Barrie Axtell
IT Consultants
Rugby, UK

Introduction

THE PRESERVATION OF food and crops by drying remains the most commonly used method worldwide. While the range of different foods that are dried is enormous, they can be divided broadly into two groups. The first includes low value foods that are dried in large quantities such as grains and pulses. Little or no value is added to these foods by drying them, the objective mainly being to provide food security. The second group includes higher value foods that are usually dried in small amounts, generally with considerable adding of value. It is this second area that holds the greatest potential for women producers to increase their incomes.

This source book has been written for workers who have little or no technical background in the preservation of foods by drying. It aims to provide a basic introduction to the principles of drying and to give project workers some understanding of what is taking place during drying, the importance of local climatic conditions and the need for proper protection of the final dry food. The all-important socio-economic context, which any technical change must consider if a project is to be successful and self-sustainable, is discussed.

This particular source book by definition is of a somewhat general nature dealing with a preservation technology rather than particular products. It aims to offer advice on the appropriate use of the sun's energy, biomass (wood and other combustible matter), and fossil fuels (oil, electricity and gas) for crop and food drying. It further aims to provide the reader with a broad perspective of the choice of drying technologies available, together with the basic principles of drying. It is therefore strongly recommended that it should be read in conjunction with other books in the series covering the processing of fruit and vegetables, grain, fish, and root crops. These outline the range of products, including dried ones, that may be prepared from such commodities.

It should be pointed out that low-moisture foods such as biscuits and snacks which could be called 'dried' are outside of the scope of this book as they are produced by totally different methods, such as baking and frying. In addition, high technology systems such as spray, roller- and freeze-drying are not covered.

The task of preparing this source book, intended for non-technical personnel, on improvements to food drying at the village and family level has not been easy because of a lack of precise information assessing the practicality and performance of the various systems that have been proposed and tested.

Case studies have been included. However, few examples could be found to illustrate drier designs which have been implemented at the field level by women producers. Not many written accounts of these field experiences are available, and therefore contributions to this section would be greatly appreciated.

1
The technical principles of drying

The purpose of drying

FOODSTUFFS ARE DRIED when they are plentiful to make them keep longer, thereby reducing waste and preserving the food for leaner times. The drying process may also be used to improve a product.

Traditional dried products mainly consist of cereals and pulses, and to a lesser extent meat, fish, vegetables, fruit and herbs. It is important to note that all traditional dried foods have been developed to suit the particular environmental conditions of the area. Good examples are dried meat known as 'biltong', from the hot, arid areas of southern Africa, or the air-dried whole legs of meat dried in the very cool, dry mountains of Europe. Each product, and its production technology, is suited to its environment. The technologies used to dry foodstuffs in the traditional way are based only on the climate: the sun, shade, low humidity and natural air flow, and sometimes the heat of a fire.

Nowadays people all over the world have been influenced by new food-eating habits, grow non-traditional crops and have packaging systems that will protect the food, once dry, from the local climate. This often means that they wish to dry foods that are not naturally in balance with and suited to the local climate. A range of new drying technologies has had to be developed to adapt to these changes.

As well as extending the life of a foodstuff and reducing losses, drying can be used for product refinement, which involves adding value to produce such as fish, fruit or herbs by creating a dehydrated product which consumers are prepared to pay more for.

Drying has several advantages for small producers:

- Most importantly, its principles are readily understood.
- In most cases packaging costs are low, plastic bags (preferably types that give good protection against moisture loss uptake) rather than bottles and tins are used.
- The final product weight is low, reducing transport costs.

Principles of drying

Drying basically involves the removal, by evaporation, of water from the surface of the product to the surrounding air. The speed with which this takes place depends upon the air (how much there is flowing around the product, its dryness, etc.) and the actual food in question (its composition, moisture content, particle size, etc.). Air contains, and is able to absorb, water vapour. The amount of water vapour present in air is referred to as *humidity*. Absolutely dry air, with no water vapour in it, has a relative humidity (RH) of 0 per cent, while air that is saturated with water has an RH of 100 per cent. The amount of water vapour that air can absorb is greatly dependent on its temperature. Tables are available which allow one to calculate how much additional water vapour can be absorbed by air at a given temperature and RH. The most important point is that as air is heated its relative humidity falls meaning it becomes drier and it is therefore able to absorb more moisture. Therefore heating the air around the product will cause it to dry more quickly.

Table 1 shows that raising the temperature substantially increases the capacity of the air to absorb water. Increasing the rate of flow of the air will also increase the speed with which water is removed from the product being dried. Table 1 shows the 'theoretical' amount of water that the air can remove. In practice these theoretical removal rates are never reached. There are a number of reasons for this such as the efficiency with which the air mixes with the product, the nature of the product itself and so on. In practice air will be able to remove 30–50 per cent of this theoretical amount. This is sometimes referred to as the 'pick-up factor' and is a useful working guide to remember when designing driers.

Table 1. Theoretical water removal rates

Temp °C	*RH*	*Water (g) that can be removed per kg dry air*
29	90	0.6
30	50	7.0
40	28	14.5
50	15	24.0

The other main factor that controls the rate at which a food dries, as mentioned above, is the character of the material itself. The two controlling factors here are the nature of the food and its particle size. Many foods have a protective outer skin, the purpose of which is to prevent them drying out. In the cases of cereals and legumes, which are normally dried whole, little can be done about this, but the drying rate of other products can be greatly increased if they are peeled or cut up.

After the surface moisture has evaporated from a food, the rate of drying depends on the speed with which internal moisture can move to the surface. This varies from product to product, for example sugary materials tend to release moisture more slowly than starchy materials, and so dry more slowly. Also, the smaller the piece of food the shorter is the distance that internal moisture has to travel to the surface, so slicing and cutting is beneficial. When cutting up foodstuffs, attention should be given to the type of knives used. As iron knives can often cause discoloration, stainless steel knives are strongly recommended.

Drying rates are crucially important if high quality products are to be produced. The conditions of moderate temperature and high humidity inside the drier are often ideal for the growth of moulds, yeasts and bacteria. From this point of view, the shorter the drying time the better. However, with some materials, particularly starchy ones, a condition known as 'case-hardening' occurs if drying rates are too fast. Case-hardening takes place when the rate of water removal from the surface is much faster than the rate at which water can move from the interior of the food to the surface.

The surface dries out into a hard layer that actually stops the migrating water reaching the surface, and so drying all but stops. In other cases, raising the temperature to increase the drying rate will destroy vitamins, result in the loss of colour and flavour, damage proteins and cause cracking of grains such as rice.

Discoloration during preparation and drying, commonly called 'browning', is caused by chemical or biochemical reactions or overheating. Some browning is the result of chemical reactions largely between sugars and proteins. Such browning is sometimes necessary in the production of certain good-quality products.

Table 2. Optimum conditions for drying selected products

Product	*Initial moisture (%)*	*Final moisture (%)*	*Drying temp (°C)*	*Pre-treatment*
Maize	35	15	60	–
Carrots	70	5	65	Blanch
Apricots	70–80	12–20	55	Sulphur
Herbs	80	5	55	–
Desiccated coconut	50	3	60	Sulphur

Well-known examples are the browning of the crust in a loaf of bread and the production of colour in certain dried fruit such as raisins.

Biochemical browning is caused by the release of enzymes from plant cells and their subsequent reaction with other natural chemicals in the material. Examples are the darkening of freshly cut slices of potato or apple. Such discoloration is to be avoided. Two common methods used are blanching in hot water or steam and the use of sulphur dioxide. These are described in more detail in the *Fruit and Vegetable Processing* and *Root Crop Processing* source books. In summary, the production of a high quality product by drying involves striking a careful balance between:

- the maximum drying rate for economic efficiency and microbiological quality;
- the minimum loss of essential components of the food;
- considering the way the dry food will reabsorb water when used.

Table 2 shows suggested drying conditions for a number of products.

Packaging of the dried product

In almost all cases, dried products, after leaving the drier, tend to absorb moisture and so should be packed in moisture-proof material as soon as they have cooled. The amount of moisture that a food can pick up depends on both the product and the climate. For example, salt picks up a lot of moisture in a humid climate and will not pour, while pepper, which picks up little moisture, flows easily. The pepper and salt have come into equilibrium with the atmosphere around them, but at different moisture levels. Every direct commodity behaves in this way, and some approximate moisture contents for a few foods, at two different relative humidities are shown in Table 3.

Table 3. Effect of relative humidity

Product	*Moisture content (%) in air of 40% RH*	*Moisture content (%) in air of 70% RH*
Tea	5	10
Coffee	9	14
Wheat	9	14
Dates	10	23

The choice of packaging material used for dry foods must take into account both the nature of the particular product and the local climate. Products which have a high capacity to absorb water clearly need a greater degree of protection.

It should be noted that the growth of micro-organisms, particularly moulds and yeasts in dried foods, depends on the nature of the food and the available moisture in it. This means that a dried fruit such as dates can safely have a moisture level of 25 per cent, whereas tea at the same moisture level would quickly go mouldy.

Those concerned with the packaging of dried foods should refer to appropriate books and take advice from specialists in this field.

Socio-economic considerations

If women are to invest in improved drying technologies, they must regard the activity as one which fulfils their needs, however those are perceived, and which provides worthwhile monetary and/or non-monetary benefits. The benefits obtainable reflect a basic distinction between two types of drier application: those of product preservation and product refinement. Product preservation is usually concerned with pulses, cereals, etc., and the main concern is to prevent food losses, with little or no value being added to the product. The amount of moisture removed in product preservation is generally not high, the crop having partially dried in the field already, and final drying taking place in and around the home. As no increase in value is achieved except through loss prevention, the actual losses in these, usually traditional, drying systems will determine the maximum cost that can be incurred if improved drying methods are adopted.

On the other hand, product refinement involves adding value, and usually involves dehydration – the removal of large amounts of water. The drying of fish, fruit, vegetables and herbs are typical examples of product refinement. This is where value is being added, and the cost of drying must be measured together with all other inputs required to produce and possibly market the product. In these cases, small-scale improved drying techniques are most likely to be accepted and adopted.

Product preservation

It is often alleged that traditional drying methods result in high-post harvest losses owing to moisture related deterioration of food. Losses occur in three ways:

- a high moisture content, which encourages attack by insects and micro-organisms;
- chemical changes which lead to loss of produce and deterioration in quality;
- physical losses owing to further processing being carried out when the produce is too wet or too dry.

Traditional drying is usually carried out in the open air, requiring consistent sunshine. Losses result from bad weather leading to deterioration of the produce, from animals eating the produce, or from contamination or infestation. There is little reliable evidence on the extent of such losses, but it has been customarily assumed that the benefits of improved drying techniques will outweigh the costs.

The introduction of small-scale driers for product preservation, however, has been less than encouraging for two reasons. First, the level of losses resulting from traditional drying practices has often been exaggerated, leading to overestimation of the benefits of any improvements (Russell, 1980; Greeley, 1986). Second, even when physical losses in foodstuffs are known, the value of benefits deriving from loss prevention is often overestimated. In particular, produce of improved quality may not attract a higher price in the local market. Also, local perceptions of 'quality' may differ from external perceptions – in south India, rice that has been attacked by insects is thought to be of higher quality because it is older and thus better tasting.

A further problem with drier services is that they cost money to use, and if people hold back from using a drier hoping for a change in the weather so that they can use traditional methods, then foodstuffs will begin to deteriorate anyway. Driers that are subject to seasonal variations in demand may not be worth the investment. In some cases, farmers have developed their own strategies for coping with poor weather at harvest time (such as mixing wet paddy with dry ones) and this helps to reduce losses.

Where dried produce is for home consumption, there are few if any cash benefits to be obtained from improved product quality. Poorer quality produce will be eaten anyway. Also, if improvements in quality are not obvious (for example, less contamination by toxin-producing microflora), the produce will not attract a higher price locally.

From a social perspective, it is worth noting that more nutritious food does not always fetch a higher price. For example, well-polished rice containing less germ is considered to be a more desirable product, although nutritionally inferior.

Driers for product preservation need land on which they can be placed and labour time on the part of their operators. It is difficult to measure these costs, but they do affect people's perceptions of the value of an improvement.

All the above considerations affect people's willingness and ability to take up an improved small-scale drying technique. In the past, focusing on product quality has led to the circumstances of people's lives which establish their priorities, being ignored.

Product refinement

Commonly, refinement means that a dehydrated product is being produced, although there are some examples of drying other refined products, drying paddy after parboiling for example. Attempts to introduce small-scale driers for product refinement have usually been part of fruit- or fish-drying projects. Often, this has been part of an income-generation programme for the rural poor.

In some cases, the drying project has involved the introduction of either new products or new marketing arrangements for existing products for which there is only a small market. Thus, while the drying unit may be a central part of the production process, the success of the overall project has depended on the development of a market. In this respect, such drying projects parallel handicraft projects in requiring a wider range of skills from project personnel, and ultimately (if successful) from project beneficiaries, to develop the market.

In other cases, the objective of projects in this area has been to improve the quality of dried products by avoiding the contamination associated with existing methods of traditional open sun-drying. However, even when the simpler solar drying methods have been used (e.g. by Mennonite Central Committee in Bangladesh for dried coconut and fish) the produce has often not been successful in local rural markets because customers have not been willing to pay any premium for the better quality product from the improved drying practice. Consequently, these dried products have to be aimed at an urban market, large-scale caterers or middle-class consumers.

Thus, in both cases the choice of drying technique is only one, and perhaps not the most central, component of the drying project. Packaging and marketing arrangements will be crucial in determining demand and it is these arrangements which have most frequently been problematic. Project success depends not only upon the organization of rural skills, resources and labour, but must also be complemented by urban trading skills. In practice, therefore, market development often means rural producers being dependent upon marketing agencies over which they exercise little influence.

Adoption of improved drying techniques

The capacity of women to adopt or adapt improved drying techniques for either product preservation or product refinement purposes will also depend on a number of social factors which shape their priorities and influence the decisions which they take. These factors may vary from place to place and overlap with each other, but are discussed in the outline below.

Time
Women's many domestic responsibilities mean that they may have little spare time, which affects their capacity to get involved in the more organized production processes that product refinement drier applications demand. They may have time to spare at some times of the year but not at others, and this will similarly limit their capacity to own and run more expensive drying equipment that has to be operated on a year-round basis. For some women, shortening the overall time that a drying activity takes will count as an improvement, even if the finished product is of the same quality. Other women may want an improved quality product that will not require extra time being spent by them.

In other cases, flexibility of time use may be more important, so that drying activities can be undertaken at the same time as other tasks. Hence the most appropriate drying technology may be an improvement to traditional techniques that shortens overall drying time, or that improves product quality without increasing time spent or one that permits time to be used more flexibly.

Family responsibilities
Women's available time and ability to become involved in drying operations will also be affected by the family responsibilities that they have. Younger women without children, and older women whose children have grown may be more able to travel to work or to work a regular shift pattern than those who have small children or obligations to their own or their husband's families. Women living in towns may have less time-consuming domestic obligations and be more ready to consider regular employment.

Credit
Access to credit is also an important factor. Where value is being added, product refinement equipment is likely to be more expensive, and require access to credit for initial purchase. Access to credit is often dependent on where people live (banks and lending institutions may not reach into rural areas), education levels (illiterate people find it harder to borrow money), and on being able to raise collateral. Wealthier women may be able to overcome some of these obstacles, but poorer women may need help with literacy and numeracy, and support in forming some kind of group organization, in order to secure a loan.

Organization
Individual women often do not have the time, the knowledge or the money to

organize a business around improved drying techniques. An organized group of women may, however, be able to share these types of responsibility between them, using other members of the group to compensate for times when certain individuals are not available. Availability of potential group members, and of support to group formation, both have to be considered.

Skills and training
Women will often have drying skills and knowledge derived from their involvement in drying of traditional products of preservation. Improvements of product preservation techniques will usually build on traditional knowledge, but it is important that, where possible, drying techniques for product refinement also recognize and build on women's existing skills. Women working with small-scale driers for product refinement need both direct technical skills, for operation and maintenance of machinery and treatment and quality control of the product, and the associated skills of literacy, numeracy, accounting, business management and marketing. They may also require training in personal skills, for example to build confidence or assertiveness. One set of skills cannot work without the others. Training courses should cover all the necessary areas, and the content should be designed in consultation with the women. Participation of women will also be encouraged if, where possible, childcare facilities are provided, courses are locally based, conducted in the local language, and are as short as is practicable.

Planning, monitoring and evaluation

Conventional economic calculations will provide an estimate of whether or not product refinement drying equipment can be used in a viable and profitable manner. Simple quantitative objectives and indicators can be derived from these calculations. However, the planning, monitoring and evaluation of benefits for all types of drier applications should take into account the women's own perspective, and their own assessments of possible risks, costs and benefits in order to obtain a more accurate impression of the impact of improved equipment or techniques upon their lives.

Planning
Plans and objectives should be developed with the full participation of the women themselves. This way the women will be able to identify their own priorities and build their own perceptions of constraints (such as lack of time or access to credit). The women should be encouraged to set technical, social and economic objectives which reflect the risks that they are prepared to take.

Needs assessment
Women should identify existing problems, with the technology that they are already using (for example, high losses), and in other areas of their lives (for example, food shortages, lack of time, lack of credit). Any training or technological inputs should be based around these identified needs, and objectives should be set according to the women's own priorities.

Potential demand/market
Project staff and technology users should establish (or decide how to establish) the potential level of demand for the dried foodstuffs, taking into account seasonality and the value added to dried products. The benefits of any improvement in domestic production should be firmly based on the women's own calculations of the value of these improvements.

Organization / business / marketing
Staff and users should discuss fully the organization of production and enterprise activities, and identify training needs where necessary. The marketing of products will often require special consideration.

Monitoring and evaluation
Having set technical, social and economic objectives, users should be encouraged to measure their progress against them during the project. Areas of interest may be technical skills, innovative capacity, time, income, access to credit, health, food security, level of education and confidence.

The checklists in Chapter 5 are intended to raise awareness of a number of factors that should be taken into account when considering improved drier application for women. They should be used as an aid to the identification of the specific needs that women might have in any particular case.

2
Traditional drying methods

IN PRACTICALLY ALL warm, sunny areas of the world, food products can be seen spread out to dry in gardens and backyards, on mats, on rocks, and hanging from the eaves of houses. Traditional methods of drying foods are also used in colder countries where there are suitable climatic conditions of low humidity. Meats and fish are still traditionally dried in Scandinavia and in the mountains of Spain, using the wind, sun and shade.

Sun drying

Simple sun drying remains the most commonly used method worldwide. In some countries, crops are often spread on roads, beaches and the roofs of houses to dry, taking advantage of the heat absorbed by the surface. Sometimes beds of flat rocks are used for the same purpose. In many

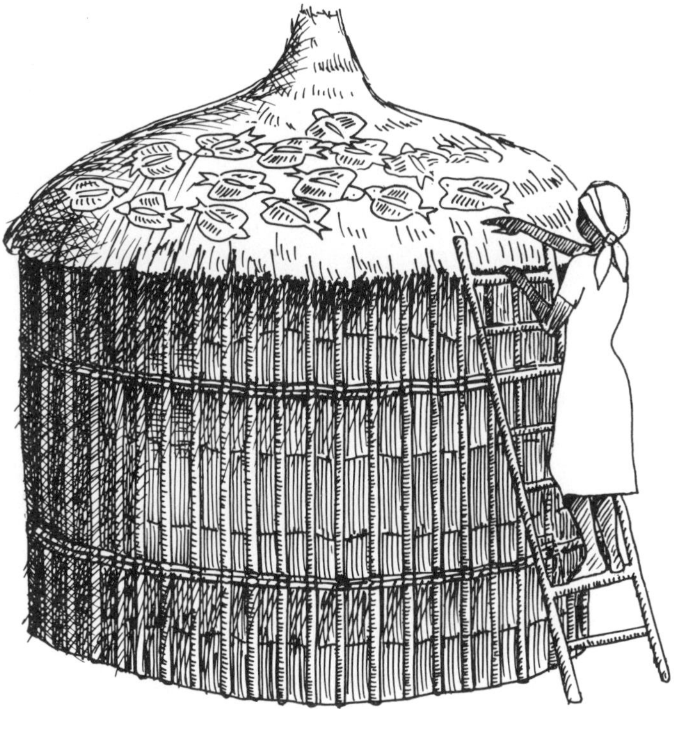

Hut roof drying

Sun drying fish on mats

cases the material is laid on mats, which reduces contamination from dirt and also facilitates easy handling.

Such simple sun drying methods have inherent advantages:

- Virtually no costs are involved as no fuel is used.
- Because permanent structures are not used, the land becomes available for farming and other uses after the drying season.

However, they also have many limitations:

- Moisture loss can be intermittent and is dependent on good weather.
- Drying rates are low and often the product will not dry fully in one day and so has to stand overnight to be finished off the next day. This increases the risk of spoilage, particularly from mould growth.
- Final moisture levels are often not sufficiently low which can lead to deterioration in storage. On other occasions, overdrying can occur.
- The product is liable to be contaminated by dust and dirt and is open to infestation by insects.
- Theft and damage by birds and animals may occur.
- In the case of bulk crops, such as grains, large areas of land are needed to spread the crop.
- More labour is required than might be expected to spread, turn and bring in the crop if rain is likely.
- Product darkening may occur and the level of certain nutrients, particularly vitamins, may be reduced by direct exposure to the sun (this is more important for some commodities than others).

Simple sun drying is applied to a wide range of commodities including fish, meat, cereals, pulses, fruit, vegetables and root crops.

Drying in the shade of eaves – Nepal

Shade drying

In certain countries, particularly those with arid climates and windy periods, some of the shortcomings listed above are overcome by shade drying using the eaves of houses, balconies and purpose-built sheds. Fading, discoloration and vitamin loss are reduced and the product is given some protection from rain showers. Shade drying is slower than direct sun drying and therefore can result in increased levels of mould growth.

Common examples of shade drying include: leafy vegetables, some herbs and spices, and the final drying of maize.

Smoke kilns

In some countries, drying is carried out in smoke kilns. This is a common practice for the production of copra from coconuts and for smoke-dried fish. In the case of coconut, an open grill constructed of slats of wood, each about one centimetre apart, is supported above a low fire. Once again, the combination of heat and smoke dries and preserves the product.

It should be noted that there is an increasing tendency away from smoke-kiln drying, partly because of concerns over the health hazard of smoke-tars in the food. In the case of coconut, such contamination passes right through the process to both the oil and residual cake.

Cooking fires

The heat of cooking fires is also commonly used to dry foods such as maize, chillies and herbs, and also to keep previously-dried foods in good condition. A store is often built above the fire either in the roof of the house or in a specially designed attic. The combined effect of heat and smoke not only maintains the product's dryness, but also acts as a repellent to insects and so reduces losses through infestation.

While these examples of traditional drying systems may appear very simple and inefficient, within the limitations, resources and demands of people in many rural areas of the world, they work effectively and have done so for thousands of years. When considered within the context of women's socio-economic position, in many cases they provide the most appropriate solution for rural producers.

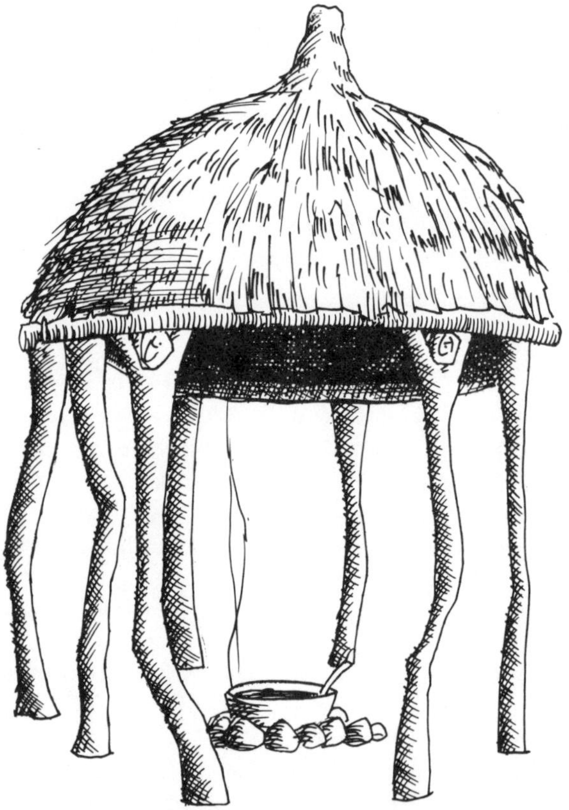

Smoke house/attic – Kenya

3
Improved drying technologies

WHEN CONSIDERING IMPROVED drying technologies it should be always remembered that, while at first sight they may provide a technical improvement, a whole range of other factors will affect their acceptance by rural women producers, who may place emphasis on aspects such as the additional time involved; the extra costs incurred; not being able to carry out drying at or near their homes; and any new skills they may have to learn.

Improvements to the traditional technologies described in the previous chapter fall into two broad groups:

- those that rely on the energy of the sun, generally termed solar driers;
- those that involve the combustion of a fuel, with or without fans, to increase air circulation.

The latter are often termed 'artificial' or 'mechanical' driers. The distinction between the two is not always clear; some solar driers may be fitted with an electric fan, or a fuelled mechanical drier may also have the facility to use solar energy in order to reduce fuel costs.

Solar drying methods, like sun drying, depend on the sun for the source of energy, but also involve the use of some form of structure to collect and enhance the sun's heat. Solar or solar-assisted drying, which reduces fuel costs, can be a sensible option for farmers, particularly when it assists or replaces artificial driers (Trim, 1982).

Solar driers

Using solar driers has several advantages over sun drying:

- They generate higher air temperatures and consequently lower humidities, which result in faster drying rates and lower final moisture contents.
- The higher temperatures generated act as a deterrent to insect and mould growth.
- The product is protected from dust and insects within the drier.
- Drying is quicker, and if racking can be used less land is needed for spreading the crop.
- They offer a considerable degree of protection from rain, which reduces the need for labour to bring in the material.
- They are comparatively cheap to build and do not require skilled labour.

Broadly, there are two types of solar driers: direct solar driers and indirect solar driers.

Direct solar driers

In such driers, the air is heated in the drying chamber which acts as both the solar collector and the drier. Perhaps the best-known type of direct solar drier is the Brace or Lawand type, shown on page 20. The sun's radiation passes through the transparent drier roof, usually glazed with plastic sheeting, or occasionally glass, and heats the drier chamber which should ideally be painted black to absorb the maximum amount of heat. The heated air then rises and leaves the chamber through the exit holes in the upper part of the back wall, being replaced by cold air entering through the entry holes in the drier base. An air flow is thus established which, combined with the reduced RH of the heated air, removes moisture from the product. To achieve maximum efficiency, the drier cover should be 'double-glazed' to reduce heat loss. Heat losses through the wood walls of driers are low but many workers consider insulated walls an added advantage.

Indirect solar driers

An indirect drier comprises two parts: a solar collector and the drying chamber. The solar collector receives the sun's radiation and is connected to the drying chamber containing the crop. A typical indirect drier is shown on page 30. Air enters the collector where it is heated. Its humidity is reduced and the hot air rises to the drying chamber by natural convection.

Such driers must be adapted to suit local climatic conditions, the crops to be dried and available construction materials.

Many solar drying projects may have failed owing to a lack of sufficient on-site climatic data. It is very important to obtain information on seasonal and daily variations of sunshine, humidity, temperature, wind speed and direction during the intended drying period.

If the crop is affected by exposure to direct sunlight, which can cause darkening and loss of sensitive components such as vitamins, it can be shaded in the drier. This is often achieved by suspending a sheet of black-painted galvanized roof sheet above the foodstuff inside the drier.

In areas of high humidity it may be necessary to increase air flows to obtain better drying rates. This can be done by fixing a black chimney to the air exit end of the drier. A chimney thus painted causes more 'draught' and hence a higher air-flow rate.

Some driers, particularly of the indirect type, incorporate large amounts of dark-coloured rocks in the collector. These, after being heated in the sun all day, continue to give off heat after nightfall, and thus the crop continues to dry.

The angle of inclination, or tilting, of the drier roof or collector is critical in order to maximize the collection of the sun's energy. The angle of the sun's rays varies between summer and winter, so when building driers, consideration must be given to the time of year when the crop is harvested. It is important in areas of rain that the roof-angle is at least 15° to allow water runoff. The example in Table 4 shows the best collector angle and direction in summer and winter for two sites, north and south of the equator (from Trim, 1982).

Table 4. Best solar collector positions

Khartoum	April	slope 5°	facing south
	October	slope 25°	facing north
Lusaka	April	slope 25°	facing north
	October	slope 5°	facing south

In this example, to allow for rainwater runoff, a good compromise angle would be 15°C and the inclined roof would face the sun. The siting of driers away from the shadow of trees, out of strong winds, etc., might appear obvious, but is often overlooked. A wind-break can be used to prevent over-cooling or physical damage.

Special plastic films for use in solar driers are not available at village-level but polythene is generally found in towns. Polythene has a short life in solar driers due to yellowing and tearing. Newer plastics which are both stronger and stable in the heat of the sun are currently available in only a few African, Asian or Latin American countries. These include:

- horticultural UVI (ultra-violet inhibited)
- polyvinyl fluorides (PVF) such as ICI Melinex and E.I. DuPont Tedlar
- polymethylmethacrylate (PMMA)
- polycarbonate (PC)
- glass fibre reinforced polyester (GRP)

When using such films, which are considerably more expensive than polythene, it is recommended that the collector is constructed in several small sections. Should any damage occur, only one small piece will need replacing and it reduces sagging of the film.

Rain can have a quite disastrous impact on solar drying. The immediate effect is a sudden cooling of the drier cover, often accompanied by it fogging over because of condensation. This brings the convection air flow to a halt. It can be some time before the drier begins to work again once the sun returns. Cheap, portable thatched drier-covers can quickly be placed over the drier at the first sign of rain and reduce the impact of rainfall.

It can be seen that while solar drying at first appears to be a simple technology, this is a misleading impression. Not only do all the considerations above – latitude, harvest period, local climate, the nature of the crop, etc. – have to be taken into account, but so do the complex cultural and socio-economic situations of the women end users.

Increasing pressure from rising populations, higher fuel prices, deforestation and climatic changes is leading to urgent food security needs. Solar drying has great potential to assist the rural producer in many parts of the world in the face of these changes.

A large amount of research and development is taking place, but despite this the uptake of solar drying technologies for productive purposes has been poor. In many cases, the technical specialists have not sufficiently involved extension workers and women processors in field testing, adaptation and transfer of these technologies. This has resulted in many failed projects. However, the case studies (in Chapter 6) from Bangladesh and Honduras show that if local people are closely involved with those supporting them then small solar driers can work and form the basis of viable production.

When considering solar drying, therefore, it is important to encourage the participation of all groups mentioned above and to ensure that local technical institutions are able to provide continuing and local long-term support.

Mixed driers and artificial driers

Solar driers suffer from certain major limitations. They cannot be used at night, and their efficiency declines in cloudy or rainy weather. Very often the product may not be completely dried in one day which may result in deterioration, particularly mould growth, during the night. In addition, they do not lend themselves easily to being scaled up into larger units without introducing construction problems and fragile structures.

To overcome these problems, several mixed driers have been proposed which use back-up sources of heat from a burning fuel when necessary. One of these, the McDowell drier, is described in Chapter 4. Artificial driers, which rely solely on the heat from burning wood, gas, oil or electricity and often have fans, can overcome the various shortcomings of solar driers. At the same time, however, they may introduce new problems. Advantages of artificial driers include:

- independence from weather conditions;
- greater degree of control of drying process;
- a wider range of products may be dried;
- greater capacity.

Disadvantages include:

- lower cultural acceptance than sun or solar driers initially, as concepts may be unfamiliar;
- higher production costs owing to fuel being used;
- higher investment needed;
- fuel, equipment, spares and maintenance skills may not be locally available;
- demand more rigid time schedules;
- mechanical technologies are often 'taken over' by men;
- generally more applicable for use in urban and peri-urban areas because of better access to fuel.

Despite these disadvantages, as the case study from Bangladesh shows, artificial driers can be successfully applied to women's projects in rural areas.

The selection of an appropriate drying system is a crucial step in project design and it is strongly recommended that a specialist's advice should be sought as a number of interrelated technical, economic and social factors are involved. These may include:

- Current drying practices, if any. Is simply an upgrading of traditional methods involved, or are new technologies and possibly new raw materials necessary? Current knowledge of drying may have a positive or negative effect. It may mean that those using traditional practices are fearful to risk change or, on the other hand, able to see the potential that proposed changes might have.
- Amount of material to be dried per day. This will be a major consideration on the size or number of driers required.
- Climatic conditions. A knowledge of local climate is crucial when solar driers are being considered. Data should be obtained from records or direct observation as mentioned previously. From these data, an objective opinion may be formed as to the feasibility of using solar driers.
- Nature of the material to be dried. Can the product be sun dried or does this negatively affect its quality? Is the material subject to heat damage as in the case of seed for planting? At this stage, advice should be sought on necessary pre-treatment which includes washing, disinfecting, blanching and cutting, etc., that may be required as well as recommended drying temperatures. The nature of the product, and local climate, will also dictate the type of packaging needed.
- Raw material availability. This is clearly of vital importance, particularly when larger-scale driers are considered.
- Local availability of construction materials and skills particularly when/if outside assistance is withdrawn.
- Fuel availability and cost. What fuels are locally available? If wood is involved, what environmental effects will its use have? What implications are there in terms of fuel-gathering time? How do different fuel options affect economic viability?
- Value added. All improved drying systems involve a capital investment. Where the products are for home use or better food security, the expenditure may not be seen in simple monetary terms, but as improving the quality of life for the women users. If income generation and sales are involved, added value becomes more significant. The value added by drying must cover equipment costs over a set period after paying fixed costs such as labour, fuel, and an agreed profit.

- Can a mix of products be dried, so increasing the time utilization of the drier during the year?
- Markets and quality demand. Many projects involving goods for sale fail due to marketing problems. Serious consideration should be given to the market. Is it local, distant, reliable? Are quality and packaging conditions demanded? What competition exists?

In addition, particularly when working with women, a range of cultural and social factors must be considered including division of labour, time availability, control of equipment.

When all of the points listed above have been considered, it should begin to become clear whether:

- the existing drying method is adequate, but perhaps attention is required in some other area such as raw material quality or markets;
- the existing drying system needs minor upgrading;
- based on climatic conditions, solar drying is, or is not, an option;
- due to climatic conditions, it would be necessary to use artificial driers;
- based on the amount of value being added, slightly more sophisticated forced air artificial driers are required.

4
Illustrated equipment guide

THIS SECTION DESCRIBES a range of improved drying systems. As far as possible an indication of ease of construction is given. In this context the term 'by user' indicates that the drier can be built using existing skills and does not involve special tools. It should be noted that a technique that has often been used when introducing driers is to start with a model that works, even if it is a little more expensive. This can then be gradually modified to use locally available construction materials and skills. In this way a trial-and-error phase is avoided. In addition, the producers become involved fully in design of the equipment.

In addition, the equipment guide tries to indicate the materials needed for construction. It should be borne in mind that for the villager any item not available in the village is essentially an import. In the context of this guide 'locally available' indicates that the materials are available from natural resources around the village. In many cases alternative construction materials are listed.

Finally, the guide indicates the level of operator technique needed, and refers to reports on the field application of the equipment. An indication of cost is included using three cost 'zones':

Low cost, \$10–\$100
Medium cost, \$100–\$1,000
High cost, above \$1,000.

Sun and air drying

Sun and air drying is most usually found when large volumes of a low-value product are being dried. Intermediate and high value crops, such as coffee, cocoa and fish are also commonly sun dried.

The most common improvements to sun drying involve raising the product above the ground on racks and covering it with fine netting to reduce contamination and insect damage and using a clean, hard, drying floor (usually concrete) and a movable foul-weather cover. Such foul-weather covers range from a simple plastic tent that can be put over the product, to specially designed structures on wheels which slide over the drying floor, to buildings into which the large drying trays are slid. Typical examples are illustrated.

Improvements to air drying simply involve constructing a roof over the product to protect it in times of rain. In some climatic regions with low relative humidity, properly exposed material will air dry efficiently even by night. Again, the use of netting to prevent gross contamination results in a better quality product.

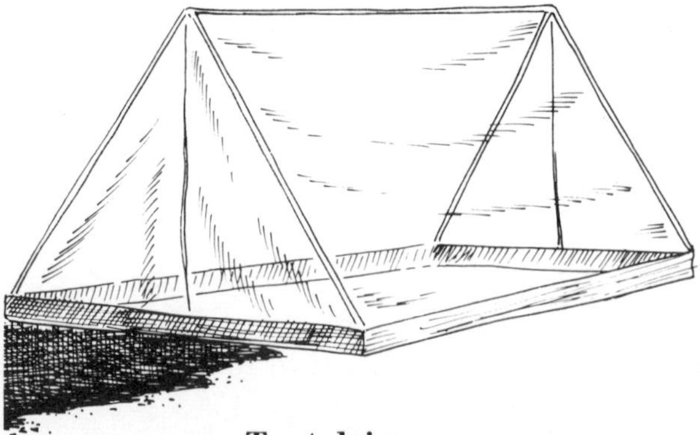

Tent drier

Sliding tray drier

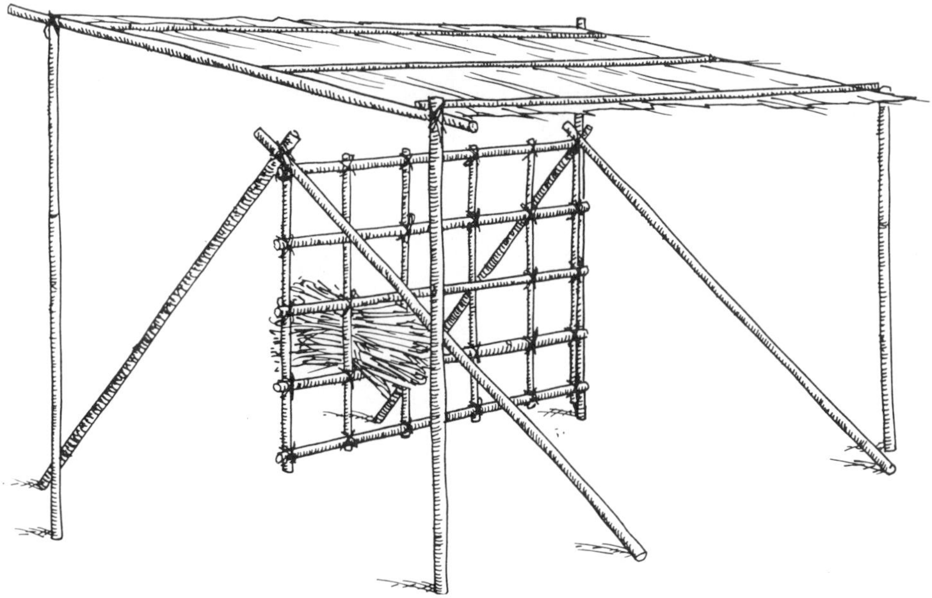

Drying under a protected roof

Direct and indirect solar driers

Brace or Lawand type cabinet drier

The basic design consists of a rectangular box some 1.8 to 2.4m by 90cm to 1.2m, with a glazed roof. The importance of using the correct angle of inclination for the collector roof, and facing north or south, has been previously mentioned. The product is supported on mesh trays. The interior should be painted black, taking care that the paint used is not toxic. To give better efficiencies, the wall and floor should be insulated, and the solar roof double glazed. In some designs the air enters through perforations in the drier base and exits through holes at the top of the walls. In other designs, as shown, the air enters through the front wall ports and exits from the rear wall.

Covering air ports with fine mesh assists in insect control.

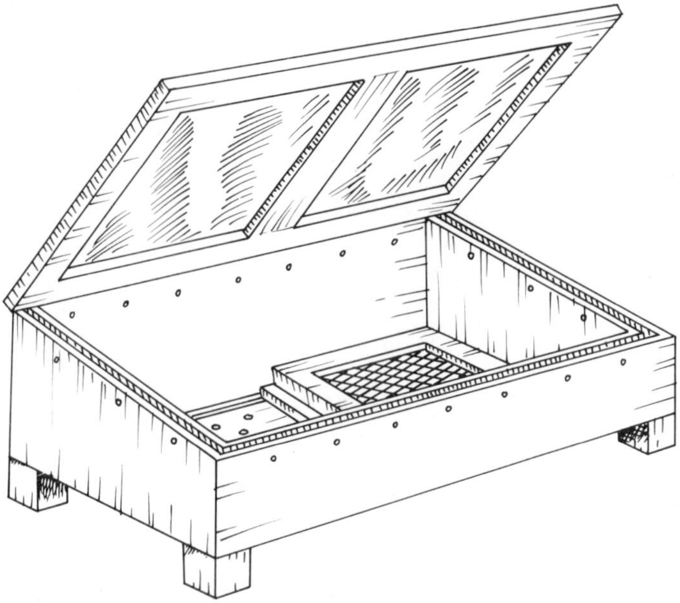

Typical Brace drier

Construction: By trained user
Materials: Can all be local apart from glazed cover. Wood, mats, mud or brick walls. Some users incorporate rice-husk for wall insulation. Trays can be made from matting, galvanized mesh or plastic mesh. Cover from polythene or UV-stable film. Clay, or wheat flour, mixed with charcoal can replace black paint.

Maintenance: By user
Operation: Simple
Cost: Low
Application: A wide range of products, such as chillies (Anon, 1980), apricots (Bhatia and Gupta, 1976), coconut (Clark, 1981), fruit and vegetables (Kapoor and Agrawal, 1973; Lawand, 1966), yams (Nahwali, 1966), fish (Trim and Curran, 1982).

A 4m by 2m unit was found to be successful for grain drying in Kenya (McDowell personal communication).

Advantages: Comparatively low cost and can be operated close to the user's home. A wide range of locally available materials can be used for construction. Provides faster and more hygienic conditions than sun drying.

Disadvantages: The presence of flies has been more noted in this type of drier than in tent or chimney driers when drying fish (described later). Slightly more expensive than tent driers. Fairly low capacity. Solar roof has short life if polythene is used rather than special plastics. Little or no control over temperatures.

Design variations: A number of modifications have been tested:

- better air distribution is said to result from internal pipes;
- placing a layer of dark-coloured stones in the cabinet base to act as a heat store can improve drying rates during cloudy or rainy periods;
- when drying vegetables, loss of colour and vitamins can be reduced if a sheet of black painted metal or black plastic is placed above the crop to shade it;
- air-flow rates have been reported to increase if a black painted chimney is attached to the drier air exit;
- covering air entry and exit holes with fine mesh reduces problems of insects entering drier;
- the use of small loading and unloading doors on rear wall avoids continually moving glazed roof and so lengthens its life.

Construction details: Step-by-step construction simply described in *Small-Scale Grain Storage Manual* (available from VITA).

Another similar design, a large walk-through drier, has been used for the solar drying of coffee in Colombia.

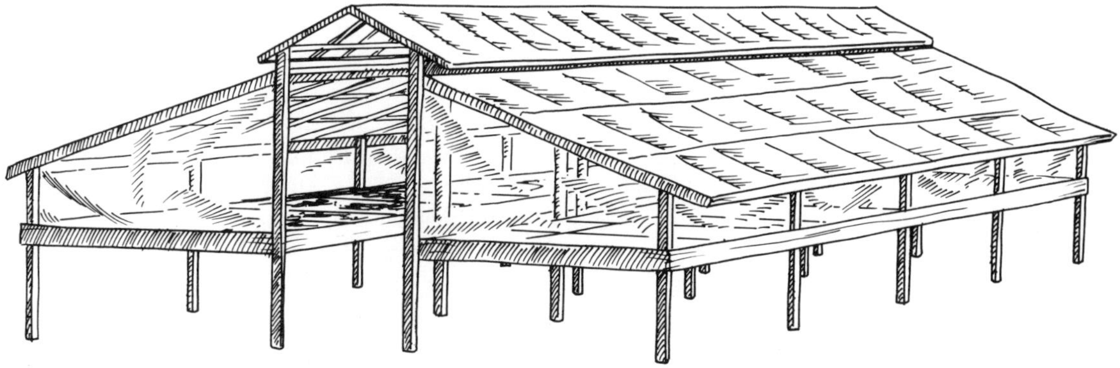

Walk-through drier

Tent drier

This is another popular and low-cost design. It essentially consists of a tent-shaped frame covered with plastic sheeting: clear plastic on the sunny side and black plastic on the shady side. Sometimes the tent may be made completely of clear plastic, but with a black floor. The material to be dried is placed on a rack at least 45cm above the ground. Access is thourgh an end flap. Air flow and temperature can be controlled by raising or lowering the bottom edge of the sides which are rolled over a pole.

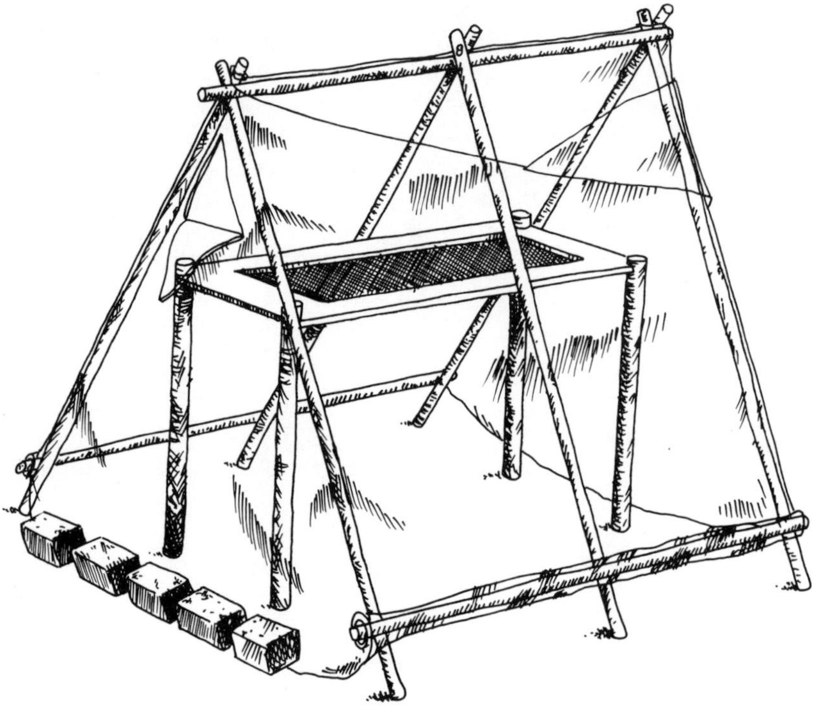

Tent drier

Construction: Simple, by trained user
Materials: Locally available wood, poles, bamboo, plastic film, etc. Polythene is generally locally available. UV stable films would need to be imported.
Maintenance: By user
Operation: Simple
Cost: Low
Application: Tent driers have been used for a number of commodities such as fruit, spices and grains. There has been particular interest in their use for fish drying. They have been used in Bangladesh (Doe *et al.*, 1977, 1979), Papua New Guinea (Anedelina, 1978), The Philippines (Pablo, 1978), Galapagos Islands (Trim and Curran, 1982).
Advantages: Provides protection, leading to better product quality, particularly related to infestation. Tent driers are cheap and easy to build and operate. Shorter drying times obtained than with sun drying, typically 25 per cent less time taken in fish drying in Bangladesh. Easily dismantled for safe storage between drying seasons.

Disadvantages: Easily damaged by wind and children who are attracted to them as play houses.
Design variations: Other tent shapes, such as semi-circular, are reported to be less liable to wind damage. In Peru a tent-like drier with several layers of trays in the shape of a rectangular box approximately 4m by 1.5m by 1m is being used for pre-cooked potato drying (see Chapter 6). Large tent driers using horticultural greenhouse designs have been used for fish.
Construction details: Should problems be encountered during construction or use, contact IT, UK.

Chimney drier

This design consists of a solar collector with a heat-absorbing blackened interior and a drying chamber fitted with a chimney. The product is supported on trays in the drying chamber. The wood-framed collector and drying chamber is covered with clear plastic sheeting. The tall chimney is covered with black plastic. The black surface absorbs more heat, thus heating the air in the chimney and causing it to rise. This gives a draught and increases the air flow in the drier.

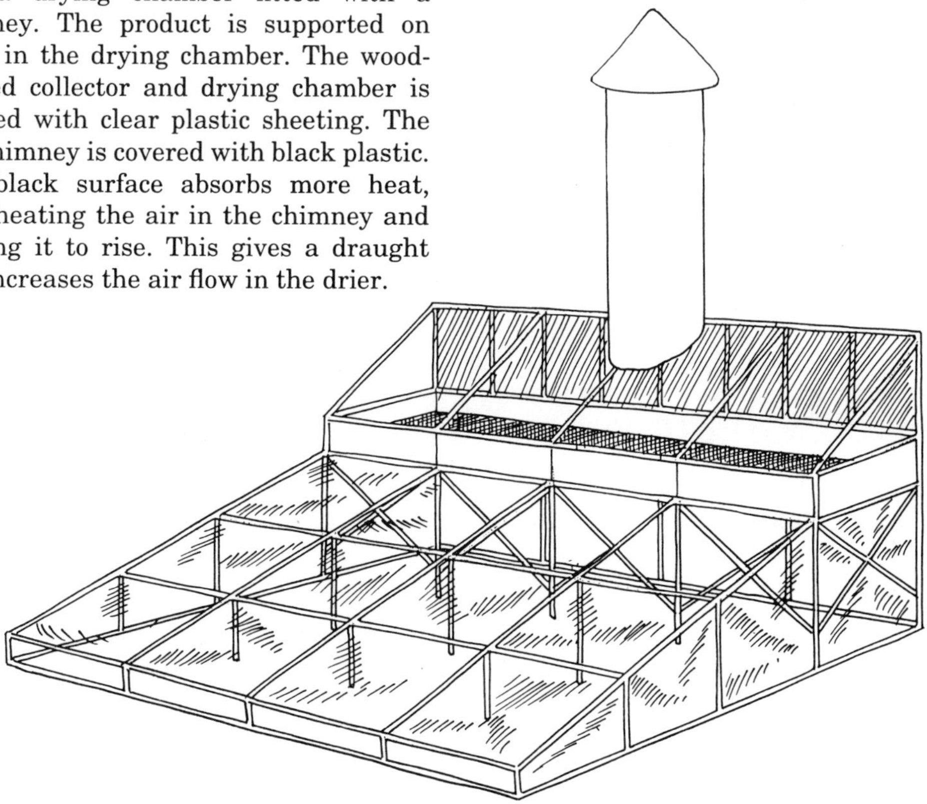

One ton indirect chimney drier

Construction: By trained user, but more complex than previous models and good training in construction required.

Materials: Locally available wood, poles, bamboo, etc. The black collector interior may be prepared from burnt rice-husk or

black plastic sheeting. Plastic sheeting to cover and black plastic sheeting for chimney.
Maintenance: By user
Operation: Simple
Cost: Medium
Application: This type of drier has been used for rice, fish, fruit and vegetables in Thailand (Boothumjinda *et al.*, 1983; Exell *et al.*, 1978, 1979, 1980).
Advantages: Can dry larger quantities of crops. For example, one ton units have been constructed in Thailand, and provide faster drying rates than sun drying in appropriate climatic conditions. Various sizes can be built – a typical Thai one ton unit had a collector of 7m by 4.5m.
Disadvantages: A semi-permanent structure, so occupying land. Subject to wind and storm damage.
Design variations: The floor of the solar collector section can be covered with dark material, such as black sand, stones or burnt rice-husk to improve heat collection efficiency.

Solar collector drier

Solar collector driers are similar to the chimney drier previously described, in that they have a separate solar collector connected to a drying chamber filled with trays. The small drier illustrated is based on a design by the New Mexico Solar Energy Association. In this design, the separate solar collector is glazed with fibreglass and contains a black painted metal sheet absorber. The wooden drying chamber contains a stack of trays. Air heated by the collector passes upwards through the trays filled with the product, and exits from the cabinet top. The feet stand in small tins containing kerosene to prevent insects climbing into the drier.

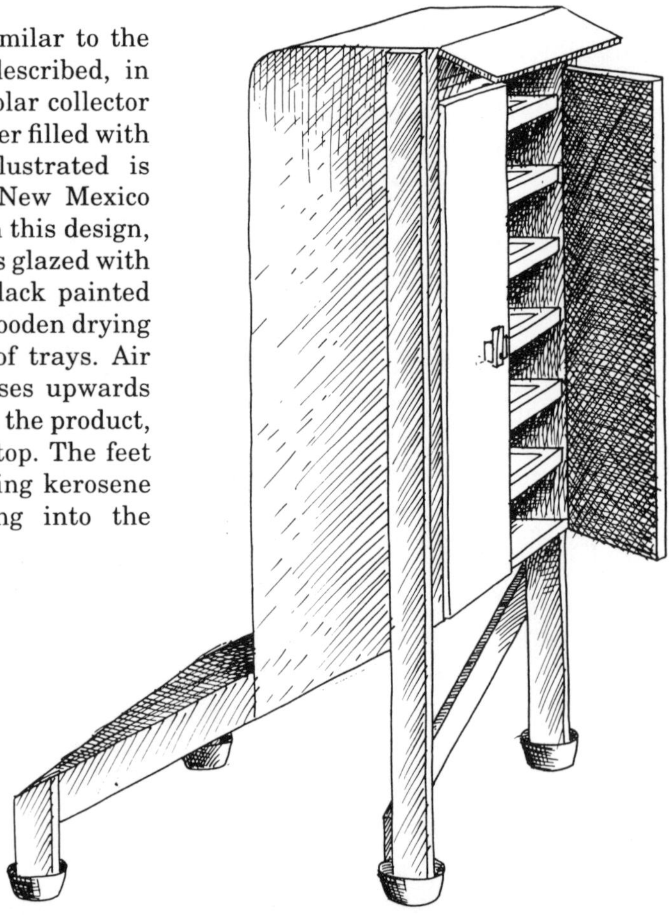

New Mexico indirect drier

Construction: Relatively simple; carpenters shop needed.
Materials: Wood, plywood, mesh for trays locally available. The fibreglass collector can be replaced by a double skin of polythene.
Maintenance: By user, but some training required.
Operation: Relatively simple
Cost: Medium
Application: Fruit and vegetables, particularly those that are affected by direct exposure to the sun. Drier successes result from use for commercial production such as on-farm preservation. This type of drier has been used for bananas in Brazil, grapes in Chile, fish in Malaysia, onions in Indonesia, taro in the USA and fruit and fish in The Philippines (Cheema and Ribiero, 1970; Gutierrez *et al.*, 1979; Martosudirjo *et al.*, 1979; Pablo, 1980, Moy *et al.*, 1980).
Advantages: Good for drying commodities that are sensitive to sunlight. Greater degree of control of temperature possible. Can dry several different types of products at the same time.
Disadvantages: Comparatively high cost for amount of product it can hold, thus more applicable to high-value commodities.
Design variations: Various driers of this type have been designed and tested. In Bangladesh, low-cost models built from woven matting, plastered with clay, and using a polythene collector have been developed for home use.
Construction details: Available from the New Mexico Solar Energy Association, Mennonite Central Committee, Bangladesh, and IT, UK.

Fuel-assisted solar biomass driers

The original design was proposed (by McDowell) to overcome drying problems in humid tropical areas. A tray supporting the product is placed under a double glazed conventional solar roof, and air enters the drying chamber through holes in the walls situated under the tray. In addition, this drier has a firebox connected to a flue that passes under the tray to an external chimney. In times of cloudy weather and/or at night, a fire can be lit and heat radiating from the pipe allows the product to continue to dry. The flue pipe passing through the drier must be smoke-proof and the flue fitted with a damper which has to be kept closed when sun drying is being carried out to avoid cooling by the flue. The firebox should be

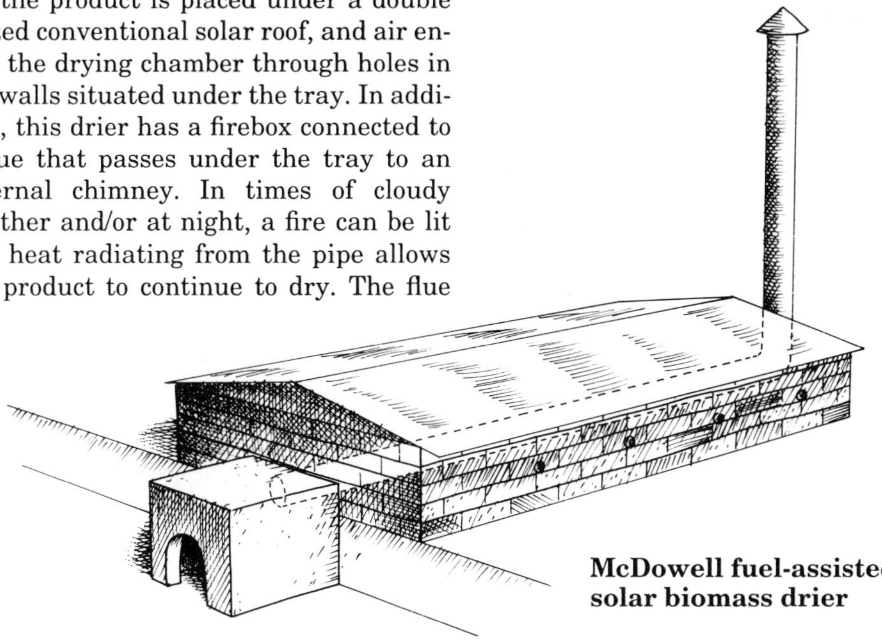

McDowell fuel-assisted solar biomass drier

facing the prevailing wind to improve draught and make sure sparks from the chimney are carried away from the polythene cover.
Construction: By trained user with assistance from metal workshop.
Materials: Brick, mud or wood walls, plastic for roof, fire-resistant hearth, metal flue pipe and chimney. All available in small towns.
Maintenance: By user
Operation: Simple, after training
Cost: Medium
Application: Fruit and vegetables, spices
Advantages: Although more costly than a solar drier of the same size, it has the great advantage of allowing the product to be completely dried in one stage. Overcomes problems of bad weather conditions. Appropriate for use in humid climates.
Disadvantages: Higher cost, some reliance on fuelwood with possible environmental effects.
Design variations: A modified design with a greater heat surface area incorporating four to six internal heat exchanger pipes connected by truck manifolds to firebox has been tested in Sri Lanka. Construction costs have been reduced by use of mud and rush-matting construction.
Construction details: Some construction details available from VITA. Matting and mud models described in *Appropriate Village Technology for Basic Services* (UNICEF, 1977).

Forced-air solar grain drier

This drier has been included as an example of a solar collector drier with the addition of a fan for forced air ventilation. Air heated in the collector is moved by a fan to the drying bin.

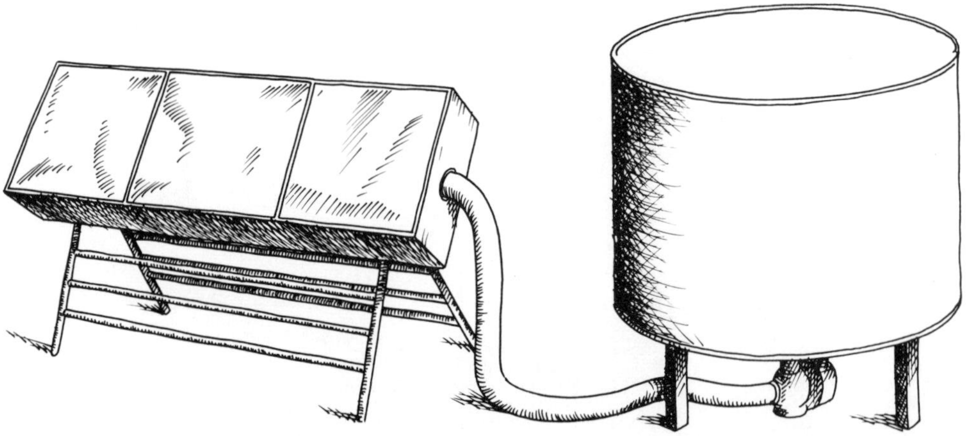

Forced-air solar grain drier

Construction: Engineering skills required.
Materials: Sheet metal, pipework, glazing materials, 0.75kW fan. All should be available in local towns.
Maintenance: Small workshop required.
Operation: Simple, after training
Cost: Medium to high
Application: This type of drier has been used in India to dry 650kg of rice to 12 per cent moisture in 10–14 hours. Other applications have included fish in India and beans in Brazil.
Advantages: Under appropriate climatic conditions, provides alternative method of bulk-drying without use of sophisticated burner systems. Very low fuel costs per tonne dried. Forced air ventilation speeds drying.
Disadvantages: Very climate dependent. Considerably more expensive than systems so far described. Degree of uptake by producers not known. Requires electricity.

Driers not dependent on solar energy

Before looking in detail at particular driers, it is important to make a distinction between direct and indirect types. In direct driers, smoke and other products of combustion pass through the product. In general this has a negative effect on final product quality.

Indirect driers, on the other hand, have some form of heat exchanger in their construction, so that only heated clean air comes into contact with the product, the smoke being led away by a flue.

Direct driers

The Ceylon kiln

This is one example of a biomass-fired direct drier. The kiln consists of a fire burning at ground level and a grill or drying platform, the whole kiln being contained in a simple house-like structure.
Construction: Simple, after training
Materials: Timber, wooden slats, thatch
Maintenance: By user
Operation: Simple
Cost: Low to medium
Application: This type of drier has been widely used to dry coconut to copra.
Advantages: Can burn low grade materials such as coconut shell which requires little or no attention after being lit. Moderate cost to construct.
Disadvantages: The resultant product is being smoked and dried, so reducing the drier's suitability for a wide range of commodities. There is a general movement away from these types of driers owing to concerns about the hazard to health posed by smoked foodstuffs. In the Ceylon kiln, the burning coconut shell requires little or no attention after being lit. Other variations using wood, etc. assume labour is available to maintain fires.
Design variations: Many country-specific variations exist, for example the Pag-

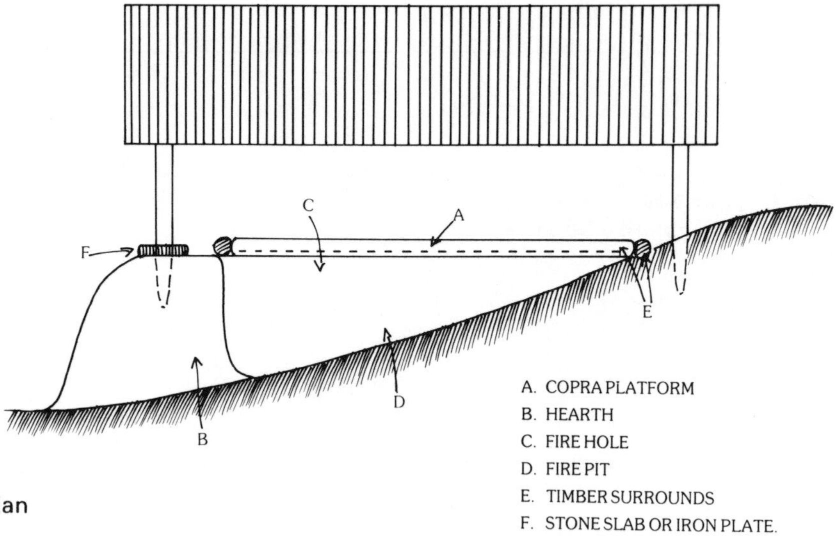

Pagsanjan

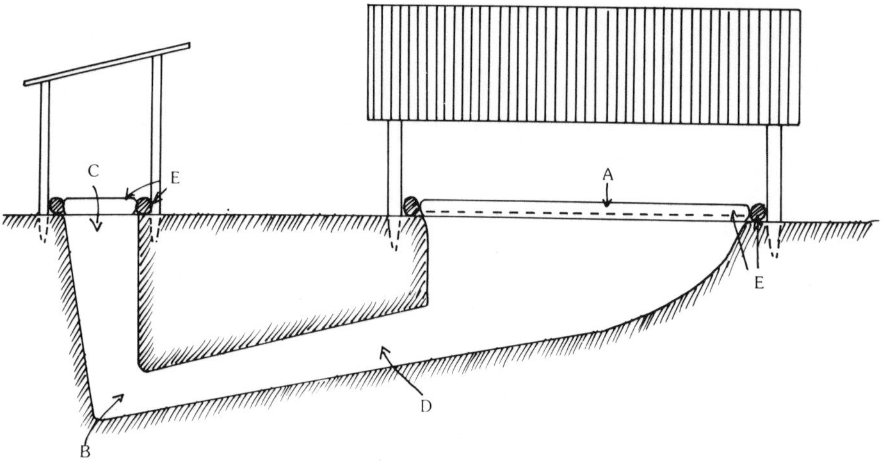

Sariaya

Pagsanjan and Sariaya direct driers, used in the Philippines

sanjan drier used in The Philippines. This is built into a hillside with a pit being excavated on one level. The higher end has an open flue, in the form of an open ditch, through which fuel is added. Very often the flue is covered with a metal plate to control the burning rate of the fire. In flat areas, a modified kiln, the Sariaya is used. It is reported to be more efficient than the Pagsanjan type as there is less tendency for the fire to be affected by wind and the hot gases are more evenly distributed throughout the product.
Construction details: Not known

Ceylon kiln

Indirect driers

In an indirect drier, the products of combustion do not come into contact with the product being dried, and this assures higher quality. However, a price has to be paid in terms of fuel-efficiency, since the heat exchanger cannot be 100 per cent efficient. Construction costs are higher and fuel consumption is greater. Therefore, indirect driers are only appropriate for larger scale production, or for processing higher value crops.

Samoan drier

Here, one or two flues or heat exchangers, sometimes arranged in a U-shaped system, are mounted under the drying platform. Chimneys are connected to the flue exits to increase draught. Sometimes the drier is fitted with a roof that can be slid on rails to allow sun drying when appropriate.
Construction: By trained user; welding of oil drums necessary.

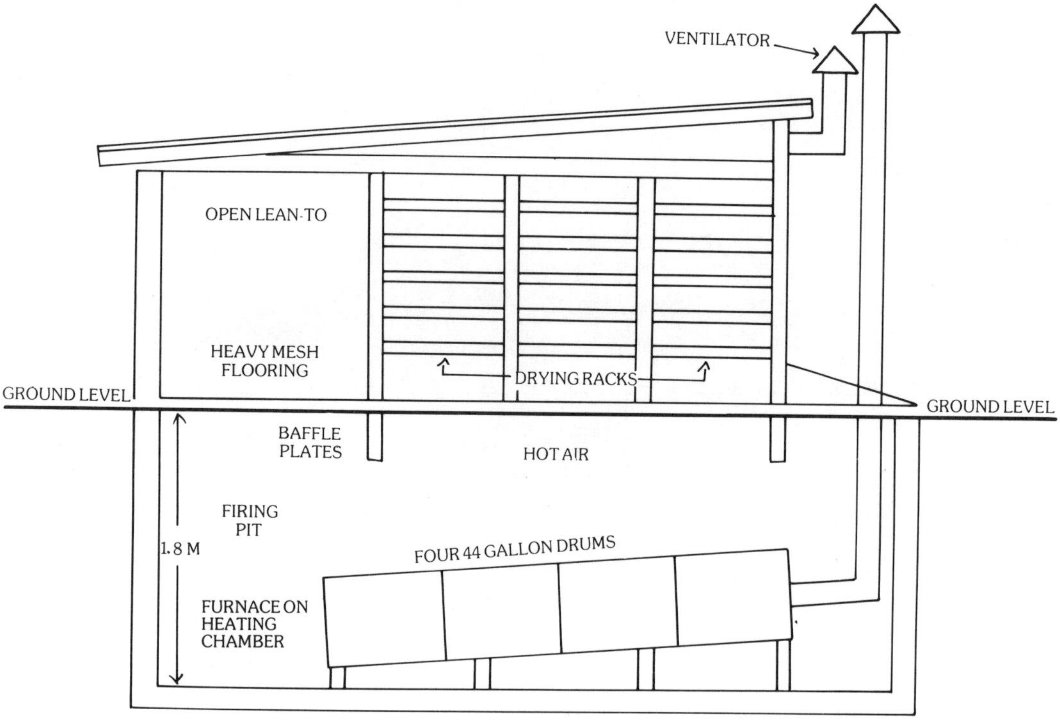

Typical Samoan drier

Materials: Wood, thatch, oil drums available locally.
Maintenance: By user (regular inspection must be carried out to check there is no corrosion of flue system.)
Operation: Simple
Cost: Medium
Application: (Papua New Guinea, Solomon Islands) coconut and cocoa.
Advantages: Reasonably cheap to construct and can produce a high quality product. Low-grade fuels such as coconut husk and cocoa pods can be used.
Disadvantages: Dependent upon adequate fuel supplies. If wood burning is involved, it may have negative environmental aspects.
Design variations: A Samoa-type drier has been used for cocoa drying (Cadbury Bros, 1963). Essentially, it consists of a fire box and simple flue in a plenum chamber below the crop, which is supported on a wooden slatted floor. Some planters have installed electric fans to increase air flow and drying rates. The Tonga dryer operates on similar principles except that the heat exchangers and firebox are placed in a pit. This involves considerable labour to construct.
Construction details: Not known

Small mechanical driers

Most driers of this type incorporate a burner, usually with a heat exchanger, and a fan blowing air through the product. A thermostat is generally fitted to control the air temperature. Common types include tray driers (where the hot air passes upwards through a series of trays bearing the product), rotary driers (which involve a rotating drum containing the product through which the air is passed) and on a larger scale (and therefore beyond the scope of this book) tunnel driers, in which small carriages containing trays pass through the tunnel. If a gas burner is used it is not normally necessary to incorporate a heat exchanger, provided the equipment is properly set up. Such powered driers, while being more costly, have several obvious advantages:

- controllability;
- independence from local climate;
- higher final product quality.

When considering systems which probably involve a major capital investment, the whole aspect of risk must be considered. Clearly, as with all systems, the value added to the product must be sufficient to cover the capital cost over the payback time, fuel costs and labour costs. In addition, aspects such as level of use, spares and servicing, and general management skills become more important.

Tray driers

A tray drier consists of a chamber stacked with trays, through which hot air is provided by some form of heater-blower usually located beneath.

There are two basic types of tray driers, batch and semi-continuous. The simplest of these is the batch drier, in which the whole chamber is filled with trays of product and air circulated until the entire charge is dry. In a semi-continuous drier, a mechanical system allows the tray, usually at the bottom of the stack (which dries first as it is closest to the hot air source) to be removed when it is dry. All the trays are then lowered and a tray of fresh raw material is entered in the space at the top of the stack. Both batch and semi-continuous systems have particular advantages and disadvantages in relation to each other.

Batch systems cost less to build and require less labour to fill and empty. Semi-continuous systems, on the other hand, are a little more expensive and complex to construct, and demand more labour to load and unload the trays periodically. They have the advantage, however, of being more fuel-efficient and producing a better quality product.

The DRY-IT batch drier

This consists of a large wooden box containing up to 24 trays. Thermostatically controlled hot air is supplied from a diesel or gas fired heater-blower and passes upwards through the trays.

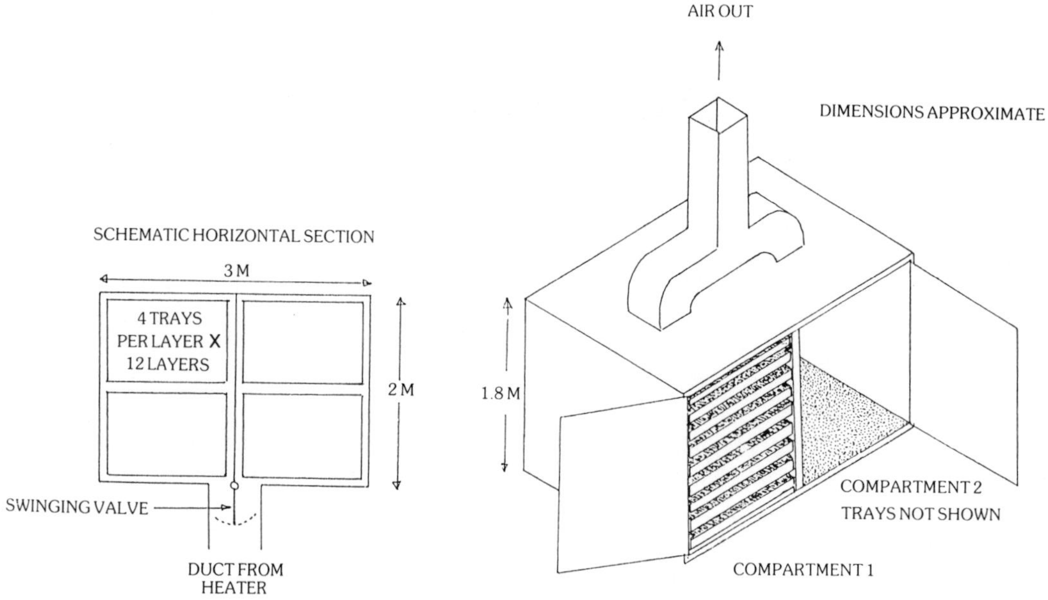

Typical double chamber batch drier

Construction: Cabinet made of wood can be locally constructed. Heater usually requires importation, although they have been built in some countries.
Materials: Wood, plywood, plastic or metal mesh for trays, heater-blower, thermostat and fuel tank. Some of these unlikely to be locally available.
Maintenance: Maintenance of the cabinet simple. Heater-blower requires workshop maintenance.
Operation: Comparatively simple, after training.
Costs: Very high.
Application: Has been applied successfully to herbs, spices, natural colouring, fruit, vegetables, desiccated coconut.
Advantages: Capacities of up to 300kg of fresh material per day possible. Low labour needed after filling. Temperature control good, resulting in high quality products.
Disadvantages: Relies on gas, kerosene and electricity. Organized working patterns necessary. Capital costs rather high, mainly due to heater-blower. More likely to be of use in urban or peri-urban areas.
Construction details: Obtainable from IT, UK.

DRY-IT semi-continuous drier

As previously described, this drier has a mechanical system allowing the removal of trays as soon as they are dry. The drying chamber can hold 16 trays, each holding 4 to 8kg of raw material, depending on the product. Throughputs of up to 400kg of raw material a day are possible.

The drier uses a similar heater to the DRY-IT batch system, and the cabinet can be constructed locally in a small, competent engineering workshop. The semi-continuous system results in higher quality products and has lower fuel consumption, but it has the great disadvantage of demanding 24 hours per day attention which in many cases might not be socially feasible. Full construction details are available from IT, UK.

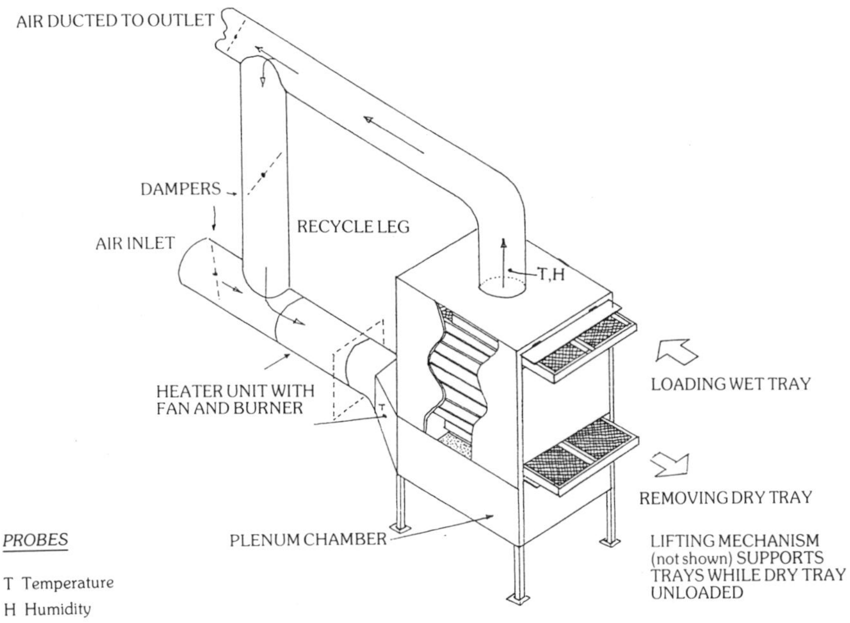

DRY-IT semi-continuous drier

IRRI paddy batch drier

The unit consists of a drying bin with a perforated floor to support the crop, into which heated air is blown from a heater-blower unit.

The bin can hold one tonne of paddy which can be dried from 28 to 14 per cent moisture in five hours. An alternative rice-husk fired furnace is available but is reported not to be as widely used. The fuel consumption per one-tonne batch is 10 litres of kerosene, plus 3.8 litres of petrol to run the small engine that drives the fan.

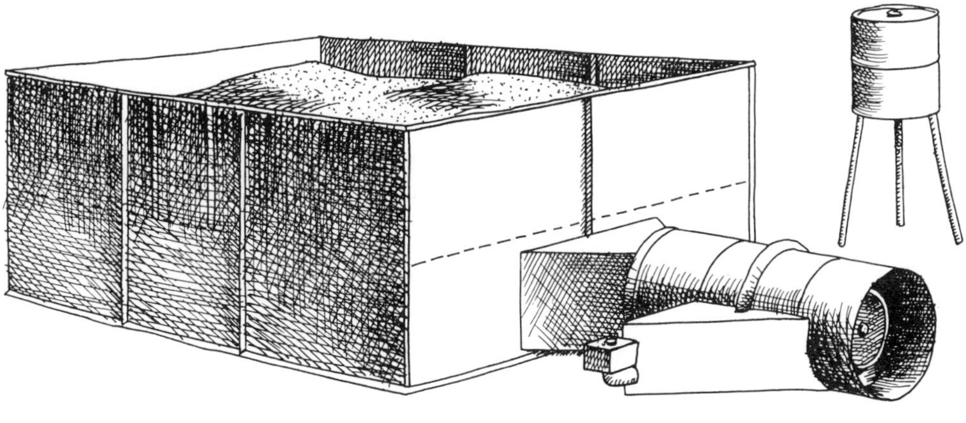

IRRI paddy batch drier

Construction: Can be constructed by medium-sized workshop.
Materials: Angle iron, plywood, special parts required for heater-blower unit.
Maintenance: Burner requires engineer. General maintenance by user.
Operation: Simple
Cost: High
Application: This type of drier has found wide use in the Philippines and in Pakistan for rice drying.
Advantages: A controllable drier, can be locally built and maintained.
Disadvantages: Requires kerosene and electricity. Relatively expensive. Limited versatility. Reported use restricted to rice.
Construction details: IRRI, the Philippines and IT, UK.

5
Project planning checklists

IT IS IMPORTANT to consider various technical and socio-economic aspects when appraising the kind of support that women might need with drying techniques. The following checklists, while not exhaustive, will help to raise some of the issues that will need consideration.

Product preservation drier applications

Are the women currently involved in drying activities?	Y ☐	N ☐
Are dried products used mainly for home consumption?	Y ☐	N ☐
Is drying usually a seasonal activity?	Y ☐	N ☐
Is the weather at harvest time usually good?	Y ☐	N ☐
Is there a market for any surplus products?	Y ☐	N ☐
Do the women have control over any income from drying activities?	Y ☐	N ☐
Does lower quality dried produce sell for similar price to higher quality produce in the market?	Y ☐	N ☐
Do the losses from traditional drying activities appear to be high?	Y ☐	N ☐
Do the women feel that losses from traditional drying are unacceptably high?	Y ☐	N ☐
Do the women have time for:		
○ learning about improved techniques?	Y ☐	N ☐
○ building improved equipment?	Y ☐	N ☐
○ spending more time in drying activities?	Y ☐	N ☐
Do the women have (or can they have) access to simple construction skills?	Y ☐	N ☐
Are other crops grown which could also be dried?	Y ☐	N ☐

If the answer to most of the above questions is 'yes', then some form of improved product preservation, drying techniques may be appropriate. It should be remembered that traditional, unimproved drying techniques may provide the best solution in cases where there are major constraints on women's time or ability to become involved in further productive activity. It is essential to consider the women's own perceptions of costs and benefits to them of improvement in traditional drying techniques.

Product refinement drier applications

Are the women currently involved in drying activities?	Y ☐	N ☐
Do the women have any experience with small enterprise activities?	Y ☐	N ☐

Have the women expressed an interest in becoming involved in small-scale drier operations?	Y ☐	N ☐
Is there a local surplus of crops for drying?	Y ☐	N ☐
Are there a number of different crops available for drying throughout the year?	Y ☐	N ☐
Will the market pay higher prices for improved quality dried products?	Y ☐	N ☐
Is the local market sizeable, or is there good transport to a sizeable market?	Y ☐	N ☐
Do the women have time and capacity to:		
○ attend training courses to learn about new equipment and techniques, and business skills?	Y ☐	N ☐
○ build improved equipment themselves?	Y ☐	N ☐
○ take on regular employment?	Y ☐	N ☐
Are the women literate and numerate?	Y ☐	N ☐
Are the women organized as a group, or do they have the optional potential to organize as a group?	Y ☐	N ☐
Will it be possible for the women to obtain or can they help to obtain credit?	Y ☐	N ☐
Will the women have control over the new processes and/or equipment?	Y ☐	N ☐
Will the women control the income that they obtain?	Y ☐	N ☐

If the answer to most of the above is 'yes', then some form of drying application for product refinement may be appropriate. It should be remembered that women becoming involved in this type of operation will need support over a wide range of activities not only concerned with use of equipment but also with running of a business. The women's own perceptions of their strength and weaknesses should form the basis of any identification of training needs.

6
Case studies

Cashew-fruit snackfood from Honduras

THIS STUDY draws on the experiences of B. Axtell of IT and M. Molina of the Institute of Nutrition for Central America and Panama (INCAP), Guatemala, to describe experiences in Honduras, in producing a dried fruit product for export.

In 1979 a small non-government organization (NGO), Pueblo a Pueblo, was set up in Honduras. The main concern initially was to assist handicraft producers who are mainly women. The organizers had a novel approach, in that they did not stop short of providing concrete marketing assistance. A strong marketing strategy was developed, resulting in the opening of a sales and distribution office in Houston, Texas.

In the early 1980s, Pueblo set out to respond to requests for help by very poor farmers, around Chuloteca in the arid south of Honduras, to process and market their cashew nuts. Some years earlier, government schemes had promoted the cultivation of cashews by means of loans, and after some five years, the trees began to yield and loans became due for repayment.

Unfortunately, no processing or purchasing system had been established to coincide with the first harvests, and it was reported that some farmers even resorted to selling trees for firewood to meet their debts.

Representatives of the INCAP visited the project and provided technical advice which eventually allowed Pueblo to train farmers in cashew nut processing and set up a production unit. The FAO publication, *Cashew Nut Processing*, proved a valuable resource and is highly recommended to those interested in this crop.

In addition to the well-known nut, cashew trees also bear a false fruit or 'apple'. This was discarded when the nuts were processed because of its bitter, unpleasant taste. It was suggested to Pueblo that this fruit might form the basis of a by-product to provide additional income and create work for the women in the community. Ideas discussed included jams, wines, vinegars and semi-crystallized dried fruit. By fortunate coincidence, the Tropical Products Institute was investigating cashew fruit utilization in Costa Rica, where there is a tradition of drying the fruit with sugar. The bitter flavour of the final product, however, made it unacceptable to many people. In 1982, an interesting paper was published on the production of a date-like caramel from cashew fruit, which involved a 'de-bittering' treatment with caustic soda followed by sugaring and drying. This information was passed to Pueblo and trials were carried out which yielded positive results.

The process is simple and involves first immersing the fruit in 1 per cent sodium hydroxide solution for three minutes, followed by thorough washing in clean water. This treatment affects the waxy surface layer of the fruit which acts as a barrier to drying and contributes to the bitter flavour. Recent communications with the producers in Honduras have revealed that the caustic soda treatment has been discontinued with no apparent negative effect on the final product's acceptability.

The fruit is then pressed between two boards to a thickness of approximately 1.5cm. This removes up to 40 per cent of the moisture in the form of juice and also makes small holes in the skin which allow quicker penetration of the sugar syrup used in later stages of the process. The squashed fruit is then simmered in a strong, hot, sugar syrup for two hours. (35kg sugar per 1000 fruit with enough water to cover.) This heating obviously contributes to good microbiological quality. The syrup can only be used a limited number of times. The cost of this key raw material must therefore be carefully considered.

After sugaring, the fruit is removed from the syrup with tongs, placed on mesh trays and dried in small Brace-type solar driers (2.4m by 1.2m).

Initially, the solar driers were not well-liked by the women involved, mainly because the polythene darkened in the sunlight and was easily damaged. IT assisted with this by providing a sample of a UV resistant film (ICI Melanex) which proved acceptable, being strong and lasting longer. Pueblo subsequently purchased 100 metres of this film, which only began to deteriorate after six years. A solar roof inclination of 9° was used, later increased to 15°, to improve efficiency. After two to three days' drying, the fruit is removed with tongs to avoid handling, and packed in units of five, in heat-sealed cellophane bags with an average net weight of 100g.

By 1986 about 40 women were involved, working in teams of 10. Each team could produce 800 dried fruits per day. The women earned 4 Lempira (US$1 = 2L) per day, which was above the local male daily rate of 3 Lempira for processing cashew nuts. The market in the USA was for 2200kg per year at $4.20 per kilo. Payback time for the solar driers was estimated to be under one week.

By 1990, the size of the group had increased to between 60 and 70 members, and 40 solar driers were in use. Sales to the US in 1990 were 4000kg and Pueblo expected to increase the size of the project for the 1991 season to meet indicated orders. A total of five more women's groups with 50 members each was planned.

This is clearly an example of a viable, sustainable income generating project. It is also one of the few examples of the successful commercial application of small solar driers.

Main points

- Small solar driers can be viable for high-value foods.
- Aggressive marketing by the NGO allows women to concentrate on production.
- UV-stabilized film is a key point in dryer acceptability.
- Communal units can meet export quality requirements.

Yachaq Mama (Power to the Women) in Peru

This group of 15 women from the Andean town of Huancayo in Peru produces a range of products including dry potato or 'papa seca'. Papa seca is a traditional product and consists of dry, pre-cooked potato pieces which are reconstituted and used in stews and other dishes.

The group formed in 1979, but it was not until 1985 that it started to process foods with the assistance of a local NGO. It now arranges for production to be carried out by two or three women in turn, thus allowing the majority to continue with other activities in the home and farm. However, all are involved in selling.

In 1988, when additional technical support became available they began to process grains and papa seca.

The initial product system used was very crude. The potatoes were hand-prepared and simply dried in the sun on mats. Production ran at only 10–15kg per month and quality was generally low. The only sales outlet was the supporting NGO's shop.

The technical support provided by a Peruvian food technologist has now helped to streamline and improve production. Output of papa seca is now at 450kg per month. The main innovations have been the use of a mechanical peeler, metabisulphite to improve the final product colour, and a simple solar drier.

The drier is essentially the tent type; rectangular in shape (4m by 1m by 1.5m), with internal racking. The climatic conditions in the high Andes are ideal for such systems: low humidity, high altitude and a good level of sunshine. The main advantages of the low-cost solar drier are the faster drying rates and greater protection from contamination.

As in many projects, the group's main problems lie not with production but in areas such as marketing, management and lack of working capital. In 1990, production rates were not consistent for a number of reasons. The process was marginally profitable, after allowing for wages, depreciation and loan repayment. This meant they could only process during the main harvest season (November to April), but not during the second crop when raw material prices were higher. Second, they were storing the final product and waiting for it to sell before producing more. As sales were low, so was production. Their lack of working capital made them fearful of entering into contracts with buyers, as they did not feel confident they could afford to meet orders.

Most of the capital equipment is now their own having repaid the loans received from the local NGO. The women believe there are greater markets but because they are not properly organized, they are not able to take advantage of such markets.

(Evaluators' final comments were: 'This group needs good training and follow-up in business management by somebody specialized in such areas, together with a new injection of working capital.')

Main points

- Small solar tend driers can achieve appreciable production rates to increase profits.
- Business and marketing aspects are as important as production considerations.
- The involvement of technical assistance substantially increased their income.

Banana drying in Thailand

This case study, considering a large, decentralized, rural banana drying operation in Thailand, was first published by the Renewable Energy Resources Information Centre, Bangkok, in 1987.

Dried banana is a major product from certain areas of Thailand, providing an annual income of 125 million Baht (US$5000). Most of it is dried by cottage industries during the October to February peak season. In Northern Thailand, a total of 50 tonnes or 150,000 bunches are dried each day during this period.

The traditional method involves direct sun drying on racks supported above the ground. Considerable losses, ranging from 50–100 per cent, can occur during the rainy season and this represents a loss of up to 3 million Baht per batch.

In an effort to reduce these heavy losses, a cabinet-style solar drier with a collector was developed by the King Mongkut Institute of Technology, Thonburi (KMITT) and field tested from 1983 to 1987. The trials showed that temperatures up to 55°C were obtained inside the drier, against ambient temperatures of 25–37°C. Drying times were reduced from between five and seven days to between four and five days and the final product was less contaminated by dust than he traditionally sun-dried product.

Each solar drier could dry about 50 bunches or 500 bananas per batch and was also suitable for household use. However, each householder processes 3000 bunches per batch and so would need 60 driers which would involve an unaffordable investment of 150,000 Baht. Other problems found were that in the rainy season the bananas could not be dried to a sufficiently low moisture content, and internal drier temperatures were not high enough to kill fruit fly larvae, which can then develop in the final product.

KMITT next looked at larger driers. The first was a twelve-times larger version of the original unit, 2m wide, 6m long and 2m high. This had a capacity of 600 bunches per batch, but the cost at 50,000 Baht was still considered too high. A solar drying hut was then designed with the same dimensions, but with a capacity of 150 bunches per batch at a cost of 3000 Baht.

In 1986, five large modified KMITT solar/collector driers and 40 solar huts were constructed. Bananas are first dried for a few days in the solar huts and final drying is carried out in the modified KMITT drier. The total capacity was 30,000 bunches per month. This system functioned well in 1986 and 1987 but then the same problems were encountered during the rainy season. Further investigation was obviously required and in 1987 most of the problems remained unsolved.

Main points

- Improved solar drier technology increased production.
- Choice of inappropriate drier resulted in unprofitable project.

Desiccated coconut production in Bangladesh: the 'Surjosnato' project

This case study of a project producing desiccated coconut, assisted by the Mennonite Central Committee (MCC), is drawn from accounts by Clark in 1981, Dirks in 1984, Martens in 1981 and Axtell *et al.* in 1991 (see references pp. 44–5). The project is particularly interesting in that it has now been running for over ten years and is one of the very few examples of the decentralized use of small-scale solar driers by women in a commercial business environment.

In mainly Muslim rural Bangladesh, women are traditionally confined to activities in and around the home. This now appears to be changing slowly, owing to increasing poverty and landlessness.

The coconut drying project started in 1977 when MCC began to investigate the solar drying of food as part of their continuing assistance programme. In its early stages, it was decided that a number of characteristics would have to be taken into consideration in designing solar driers for use by the women in Bangladesh. They should be low-cost, simple to operate, protect the product from contamination, and give reasonably rapid drying. During the early stages three types of driers were investigated; an indirectly-heated solar drier of the Exell type, a solar-box drier which had been proposed by McDowell and a solar cabinet drier known as the Brace drier (Exell *et al.*, 1978; McDowell, 1973; Lawand, 1978). Early work showed that the Exell-type drier was not suitable for the project, being more expensive than the other two designs and having poor air flow characteristics. Later designs of this drier were modified with improved air-flow rates. The McDowell-type solar box drier, although very cheap and widely used in East Africa, was unsuitable for use in Bangladesh because it failed to give internal temperatures of more than 50°C, the inside of the plastic cover fogged over owing to insufficient air flow, and the drier was not portable. This last point was important in Bangladesh as driers must be protected from flooding. However, with the use of a double-glazed cover and a strict construction specification of the air outlet, such problems were overcome.

The Brace-type cabinet drier proved to meet the project's needs most effectively. It is cheap to make, and is constructed of panels made up of a 5cm thick sandwich of woven bamboo sheets packed with rice straw as insulation. The life of the drier is approximately three years, if properly cared for, although the polythene collector had to be replaced every three to five months owing to it becoming brittle and opaque after prolonged exposure to the sun.

The project staff carried out a considerable amount of marketing work to find out which products would be most suitable for them to process. It was finally decided that desiccated coconut offered the most potential, as it is widely used in Bangladesh, but all of it was imported.

The first producer group was organized in 1980 to produce solar-dried grated coconut. Each producer was equipped with one solar cabinet drier with a capacity to dry 20 coconuts per day (depending on variety, a yield of 2–2.5kg when dried). The producers bring the dry product to a central collection point where it is graded and checked for quality, packaged by the project staff and transported to the capital, Dhaka. The product has been given the name 'Surjosnato' or rays of sunshine. After making loan repayments, each member of the group realizes an income of about $18 per month.

As the project developed marketing links with buyers, it was realized that in order to keep these customers satisfied the project would have to maintain a reliable supply throughout the whole year. The problem was of course that it was impossible to dry during the monsoon or rainy season in solar driers. MCC thus required a low-cost system for drying during wet weather, and one of their technologists visited the UK to conduct trials using the IT continuous tray drier. During these trials it became clear that the use of the IT continuous drier would be a more viable choice.

The IT continuous drier was locally modified. The heater (originally gas or diesel fired) was replaced by a coconut-husk fuelled heater. With this modification, the system was able to supply air at 90–95°C.

It was soon found that the tray drier when used on its own for drying fresh coconut was unlikely to be economic. However, its use for finishing the drying of almost-dry material from the solar dryers, and taking over in periods of rain meant that the risk of losing product due to spoilage was removed. While the financial success of the business has been mostly because of the solar driers, the artificial drier has ensured the project's continued commercial viability.

In 1984, 20 women were involved. This had grown to 100 in 1991. These women come from eight villages within a 1.5km radius and they visit the project centre about once each week to collect their coconuts. The day-to-day management of Surjosnato is carried out by a paid manager and a quality controller/supervisor. The manager is responsible for all aspects of the raw material purchase and sale of final produce. Decisions affecting the project are made by a committee consisting of two women coconut producers, two of the coir workers, the manager and the quality control supervisor, together with a representative of MCC. A producer chairs the committee.

In interviews, the women producers expressed a strong sense of control over the project and they felt they could influence the management. In general, profit margins in rural Bangladesh are small, and this is true of the coconut drying project where the profit (after paying off fixed costs and allowing for depreciation) is some 5 per cent of total sales. A substantial portion of the profit is used to pay off loans and the rest is redistributed as a dividend. In reality, the main benefit to the producers in this project is the wages they receive. The loans initially provided by MCC were interest free and it was left to the producers to decide how much to pay off, as well as when and how much to distribute in dividends. By 1990, the project members had begun their own savings bank, depositing small amounts of money with 10 per cent interest being paid.

The main problem for a commercial entrepreneur would be the decentralized arrangement of the business and high level of working capital necessary for the purchase of coconuts in bulk when prices are low. This required an investment of $7,150 in 1988.

Main points

- One hundred women work with 75 solar driers.
- The producers concentrate on processing of the coconut while the project management is responsible for organization and marketing.

Potato drying in India

In this case study Robert Nave of Compatible Technology, USA, describes his experiences with village-level potato processing in northern India. The drying system used is the most simple possible, sun drying, with only one change, the covering of the product with fine netting to provide protection.

Potato production has increased in north India, especially in western Uttar Pradesh. The increased production causes gluts at harvest time resulting in very low prices, followed (within two months) by high price increases. Only those who can afford to buy large quantities of potatoes and hold them in cold storage benefit from these price fluctuations.

Over the years, a tradition has developed of eating a deep-fried dried potato chip, the demand for which increases greatly during certain religious festivals. The demand for the snack food is rising but the dried chips have generally been of very poor quality – being made in very small quantities at home or in cottage industries using unhygienic methods.

Against this background, Compatible Technology Inc. of Minneapolis, USA (CTI) in collaboration with the Society for the Development of Appropriate Technology, Bareilly, India (SOTEC), have worked together to develop storage systems, processing equipment, an extremely simple production system and a marketing organization for dried potato products. The project received considerable support from the International Potato Centre (CIP).

The project is organized into groups of small, independently-owned village level units, called Tier One units, with ownership by an individual, a family or an association of people. They are registered as societies to give them legal status. Each is operated by 8 to 12 people and is able to process some 60 tonnes of fresh potatoes per drying season. Eleven such units are now operating, seven of which are owned by small farmers. Over half of those involved are women, and one unit is composed entirely of women. CTI is now actively encouraging even greater participation of women in the project.

Tier One units are bound together in a registered association which has all the characteristics of a co-operative. This is responsible for collecting the final product, maintaining quality control, dealing with finances, arranging the supply of equipment and chemicals as well as storing and dispatching the finished product. SOTEC and two marketing people are also members of the association. Marketing and sales are carried out by independent SOTEC companies.

After purchase, the potatoes are sorted and any poor material is immediately processed, the remainder going into store for up to three months. They are then peeled, washed and blanched for three minutes. They are rinsed well to wash away starch, then sliced and placed in a metabisulphite solution to keep the colour fresh.

The drying system is very simple. The diced potato is placed on racks off the ground, covered with fine netting and dried by the effects of sun and wind. The total equipment cost per unit is US$2000, including masonry for the blancher, netting for drying racks, tubs, slicer, peeler and plastic bags.

Upon receipt of the dried product the quality is checked and the second grade is used for processing by grinding to granules or powder. The three types of product – chips, granules and powder – are then either sold in bulk to wholesalers or, on order, packaged into retail size packs.

The project has found that the introduction of higher-quality dry chips has created a substantial increase in market demand. It is considered that the project will need to expand tenfold to meet the existing demand.

Main points

- Involves 70 to 80 women with perceived potential for tenfold expansion from present levels.
- A low-value commodity but drier costs kept to absolute minimum.
- Producers are left to produce; business and marketing aspects are the responsibility of the association.
- An improved quality traditional product opens up a wider market.

References and further reading

On drying

Anedelina, E. (1978) *Sub-Project: Inland Fisheries* Proceedings of UNESCO Solar Drying Workshop, Manila, Philippines. Bureau of Energy Development.

Anon (1965), 'How to make a Solar Cabinet Drier for Agricultural Produce'. Do-it-yourself leaflet 16, Brace Research Institute, Canada.

Anon (1968), 'Preserving Food for Drying', A Math/Science Teaching Manual. No. M-10. A very good teaching manual for people involved in education at the junior high or high school level. Describes physics of solar energy design; the physiology of dried foods; health and nutrition.

Anon (1980), 'Chilli Drying, Vegetable Seeds Drying'. Annual Report, Central Institute for Agricultural Engineering, Bhopal, India.

Axtell, B., Bush, A. and La Cruz, G. (1991), *Try drying it*. IT Publications, London, UK.

Bergen, K.T. and Bergen, I. (1985), 'More Than Just Sunshine'. Paper presented at workshop on solar drying of food and food crops, Bangladesh Agricultural University, Mymesing, Bangladesh.

Bhatia, A.K. and Gupta, S.L. (1976), 'Solar Drier for Drying Apricots'. *Research and Industry*, 21, pp. 188–91.

Boothumjinda, S., Exell, R.H.B., Rongtawng, S. and Kaewnikom, W. (1983), *Field Tests of Solar Rice Dryers in Thailand*, Proceedings of ISES Solar World Forum, pp. 1258–1263, Perth, Australia. ISES.

Byrne, J. *et al.* (1978), *The Savar Experiment*. The Lutheran World Federation, Dhaka, Bangladesh.

Chakraborty, P.K. (1976), 'Solar Drier for Drying Fish and Fishery Products'. *Research and Industry*, 21, pp. 192–4.

Cheema, L.S. and Riberio, C.M.C. (1970), 'Solar Driers of Cashew, Banana and Pineapple'. Proceedings of Conference 'The Sun: Mankind's Future Source of Energy', pp. 2075–9, International Solar Energy Society, Australia.

Clark, C.S. (1981), 'Solar Food Drying: Rural Industry', *Renewable Energy Review Journal*, Vol. 3, No. 1.

Clark, C.S. and Saha, H. (1982), 'Solar Drying of Paddy', *Renewable Energy Review Journal*, Vol. 4, No. 2, pp. 60–5.

Curran and Trim (1982), *Comparative Study of Three Solar Driers for Use with Fish*. FAO Expert consultation on Fish Technology in Africa, Casablanca, Morocco.

DANIDA (1974), *Report on DANIDA Drying Project*. Bangladesh Academy for Rural Development, Comilla, Bangladesh.

De Padua (1976), *Rice Post-Production Handling and Processing: its Significance to Agricultural Development*. Paper presented at the International Workshop on Accelerating Agricultural Development, Searca, Laguna.

de Silva, N. (1976), *Consumer Preference and Standards for Dry Fish and Other Dried Foods. Sun Drying Methodology, A Seminar Discussion Report*, pp. 30–1, National Science Council of Sri Lanka, Colombo, Sri Lanka.

Dirks, D. (1984), 'What is Solar Drying?', *Generator*, Vol. 1, No. 2, pp. 9–10.

Doe, P.E., Ahmed, M., Muslenmuddin, M., Sachithanantha, K. (1977), 'A polythene tent dryer for improved sun drying of fish', *Food Technology in Australia*, Vol. 29, pp. 437–441.

– (1979) 'A polythene test fish dryer – A progress report', Proceedings of an international conference on Agricultural Engineering in National Development, doc. 79–12, Selangor, Malaysia.

Exell, R.H.B. (1980), Basic design theory for a simple solar rice dryer', *Renewable Energy Review Journal*, Vol. 1, No. 2, pp. 1–14.

Exell, R.H.B., Kornsakoo, S. (1978), 'A low-cost solar rice dryer', *Appropriate Technology*, Vol. 5, No. 1, pp. 23–4.

– (1979), 'Solar rice dryer', *Sunworld*, Vol. 3, No. 3, p. 75.

Greeley, M. (1986), *Rice in Bangladesh: Postharvest Losses, Technology and Employment*. D.Phil thesis, University of Sussex.

Gutierrez, H.N., Alvarado, S. and Puyol, E. (1979), 'Construction and Experimental Tests of a Radiative/Convective Type of Solar Dryer'. *Simiente Investigaciones*, 49, 3–4, 51–57.

– (1980) *Design of Paddy Driers*. Paper presented at the Post–Production Workshop on Food Grains, Bangladesh Council of Scientific and Industrial Research, Dhaka, Bangladesh.

Harigopal, V. and Toniapi, K.V. (1980), 'Technology for Villages – Solar Drier', *Indian Food Packer*, Vol. 34, No. 2, pp. 48–9.

Howarths, S. (1978), *Solar Drier*. Technical Paper 34, p. 3, Paktribas Agricultural Centre, Dhankuta, Nepal.

Hurley, E.G., *et al.* (1980), *Rice Post-Harvest Technology Project: Project Report for the Two Years 1978–1979*. Bangladesh Rice Research Institute, Dhaka, Bangladesh.

ILO (1976), *Solar Drying: Practical Methods of Food Preservation*. ILO Publication, CH-1211, Geneva 22, Switzerland.

Ismail, M.S. (1980), 'The Drying of Fish and Fish Products in Malaysia'. Paper presented at Conference on Solar Energy Utilization.

Kapoor, S.G. and Agrawal, H.C. (1973), *Solar Dryers for Indian Conditions*. Conference proceedings. Paris, Unesco.

Kennedy, L., Wood, C.D. and Oswin, C.D. (1983), *The Use of Solar Driers to Reduce Losses of Sun Dried Fish During the Wet Season in Malawi*. Report of the Tropical Producers Institute.

Khan, E.U. (1974), 'The Utilization of Solar Energy', *Solar Energy*, Vol. 8, No. 1, pp. 17–22.

Lawand, J.A. (1966), 'A Solar Cabinet Drier', *Solar Energy*, Vol. 10, No. 4, pp. 158–64.

– (1978), *Proceedings of the Solar Drier Workshop Manila, Philippines*. The focus of this book is the drying of foods in humid tropical regions of the world. More technical information than practical.

McDowell, J. (1973), *Solar Drying of Crops and Food in Humid Tropical Climates*. Report CFNI-T-7-73, p. 42, Caribbean Food and Nutrition Institute, Kingston, Jamaica.

Martens, R. (1981), *A Solar Drier Applied to a Village Food Processing Industry*. ADAB News.

Martosudirjo, S., Kurisman, S. and Taragan, I. (1979), *Improvement of Solar Drying Technique in Post-Harvest Technology – A Study of Onion Drying in Indonesia*.

Proceedings of Inter-Reg. Symposium on Solar Energy for Development, Paper B-1, Tokyo, Japan.

MCC (1985), *Yearly Report on Employment Raising Programme.*

Moy, J.H., Bachman, W. and Tsai, W.V. (1980), *Solar Drying of Taro Roots.* Transactions of ASAE, 242–246.

Nahwali, M. (1966), 'The Drying of Yams with Solar Energy'. Technical report T27, Brace Research Institute, Canada.

New Mexico Solar Energy Association (no date), *How to Build a Solar Crop Drier.* 10 pp., Sante Fe, USA.

New Mexico Solar Energy Association (1978), *How to Dry Fruit and Vegetables.* AFPRO, New Delhi.

Pablo, I.G. (1978), *The Practicality of Solar Drying of Tropical Generation in Rural Areas.* Proceedings of UNESCO Solar Drying Workshop, Manila, Bureau of Energy Development, Manila, The Philippines.

– (1980), *Solar Drying of Agri- and Marine Products.* Presented at International Symposium on Recent Advances in Food and Science Technology, Taipeh, Taiwan.

Richards, A.H. (1976), *A Polythene Tent Fish Drier for Use in Papua New Guinea's Sepik River Salt Fish Industry.* Proceedings of Seminar 'Sun Drying Methods', National Science Council, Colombo, Sri Lanka.

Russell, D.G. (1980), *Socio-economic Evaluation of Grains Post-Production Loss-Reducing Systems in South East Asia.* Paper presented at the E.C. Stakman Commemorative Symposium: 'Assessment of Losses which Constrain Production and Crop Improvement in Agriculture and Forestry', University of Minnesota, Minneapolis.

Ryland, G.J. (1985), *The Economics of Grain Drying in the Humid Tropics.* Australian Centre for International Agricultural Research, Canberra, Australia.

Shaw, R. and Booth, R. (no date), *Simple Processing of Dehydrated Potatoes and Potato Starch.* International Potato Centre, Lima, Peru.

SKAT (1980), *Desecador Solar Simple.* Swiss Centre for Appropriate Technology, St Gallen, Switzerland.

Szulmayer, W. (1971), 'From Sundrying to Solar Dehydration', *I. Methods and Equipment, Food Technology in Australia,* Vol. 23, pp. 440–3.

Trim, D.S. (1982), *Solar Crop Driers.* TPI/ODA, London, UK.

Trim, D.S. and Curran, C.A. (1982), 'A Comparative Study of Solar and Sun-Drying of Fish in Ecuador'. Report L60, TDRI, London, UK.

Trim, D.S. and Ko, H.Y. (1982), 'The Development of a Forced Convection Solar Drier for Red Peppers'. *Tropical Agriculture.*

Umarov, G.G. and Itramov (1978), *Features of the Drying of Fruit and Grapes in Solar Radiation Drying Apparatus.* Gelio Tekhnika, Vol. 14, No. 6, pp. 55–7.

US Peace Corps, based on: Zweig, P., *et al.* (1980), *Improved Food Drying and Storage Training Manual.*

US Peace Corps (1980), *Solar and Energy Conserving Food Technologies: A Training Manual.* Information Collection and Exchange Office of Training and Program Support, 806 Connecticut Avenue, Washington, DC 20526, USA.

Van Arsdel, W.B., Copley, M.J. and Morgan, A. (1973), *Food Dehydration.* Vols. I and II, pp. 347 and 529, Ari Publishing Co. Inc., Westport, USA.

Watanabe, K. (1975), 'An Experimental Fish Drying and Smoking Plant on Volta Lake, Ghana: Design', *Tropical Science*, Vol. 17, No. 2, pp. 75–93.

Wereko-Brobby, C. (1985), *Solar Driers, their role in post-harvest processing*. Commonwealth Secretariat Publications, London, UK.

On management of small-scale projects

Jackelen, H. (1983), *Manual for commercial analysis of small scale projects*. AT International, Washington, USA.

International Women's Tribune Centre (1981), *Women and Small Business*. International Women's Tribune Centre, 777 UN Plaza, New York, NY 10017, USA.

MATCOM project (no date), 'Curriculum Guide for Agricultural Co-operative Management Training'. CO-OP Branch, International Labour Office, Geneva, Switzerland.

On women and food cycle technologies

Carr, M. (1982), 'Has Anything Changed for Women?', *AT Journal*, 1982, Vol. 9, No. 4.

Carr, M. (1984), *Blacksmith, Baker, Roofing-sheet Maker*. IT Publications, London, UK.

Stephens, A. (1986), *Yes, Technology is Gender Neutral, But . . .* CERES, 108, FAO, Rome, Italy.

Tinker, I. (1984), *New Technologies for Food Chain Activities*. USAID, Office of Women in Development.

On Appropriate Technology

Canadian Hunger Foundation (1976), *A Handbook on Appropriate Technology*. Canadian Hunger Foundation, 323 Chapel Street, Ottawa, Ontario K1N 7ZZ, Canada.

Carr, M. (1985), *The AT Reader: Theory and Practice in Appropriate Technology*. IT Publications, London, UK.

Carruthers, I. and Rodriguez, M. (eds) (1992, 4th edn), *Tools for Agriculture: A guide to appropriate equipment for smallholder farmers*, IT Publications, London, UK.

Darrow, K. and Pam, R. (1981), *Appropriate Technology Source Book*. Vols. 1 & 2, Volunteers in Asia, P.O. Box 4543, Stanford, California 94309, USA.

Fellows, P. and Hampton, A. (1992), *Small-scale Food Processing: A guide to appropriate equipment*. IT Publications, London, UK.

Hale, P.R. and Williams, B.D. (1977), *LIKLIK BUK: A Rural Development Handbook*. Catalogue for Papua New Guinea, Liklik Buk Information Centre, P.O. Box 1920, Lae, Papua New Guinea. This book gives community level leaders and trainers a broad range of information on crops, animals, processes, designs, health, and animating rural development.

Skill Development for Self Reliance Project (no date). Catalogue of items of Appropriate Technology accepted by people in rural areas. Specific learning packages are developed for the different items to be used as training material.

UNICEF/Kenya Go (no date), *Appropriate Village Technology for Basic Services*. ILO-SDSR, P.O. Box 60598, Nairobi, Kenya. A catalogue of devices displayed at the UNICEF/KENYA Government Village Technology Unit, Eastern Africa Regional Office, P.O. Box 44145, Nairobi, Kenya.

VITA (1970), *Village Technology Handbook*. VITA, College Campus, Schenectady, New York 12308, USA.

Contacts

Information and advice on drying food is available from the following organizations.

Africa

NHRI
National Horticultural Research Institute, IDI-ISHIN, P.M.B. 543Z, Ibadan, Nigeria.

ONORSOL
Office National de la Recherche Solaire, BP 621, Niamey, Niger.

TCC
Technology Consultancy Centre (Department of Mechanical Engineering), University of Science and Technology, Kumasi, Ghana.

Food Processing Research Centre
Khartoum, Sudan.

Department of Agricultural Engineering
University of Zambia, Lusaka, Zambia.

Appropriate Technology Centre
Kenyatta University College, P.O. Box 43844, Nairobi, Kenya.

UNICEF
United Nations Childrens Fund, Regional Office for Central and West Africa, B.P. 433, Abidjan 04, Ivory Coast.

Asia

BRAC
Bangladesh Rural Advancement Committee, 66 Mohakhali Commercial Area, Dacca–12, Bangladesh.

AFPRO
Action for Food Production, C52, ND South Extension II, New Delhi–16, India.

CFTRI
Central Food Technological Research Institute, Mysore 570-013, India.

Bandung Institute of Technology
Development Technology Centre, P.O. Box 276, Bandung, Indonesia.

MCC
Mennonite Central Committee, P.O. Box 785, Mohammadpur, Dacca 2, Bangladesh.

IRRO
International Rice Research Institute, P.O. Box 933, Manila 1099, The Philippines.

Europe

GATE/GTZ
German Appropriate Technology Exchange, Postfach 5180, D-6236 Eschborn 1, Germany.

SKAT
Swiss Centre for Appropriate Technology, Varnbuelstrasse 14, St Gallen CH 9000, Switzerland.

GRET
Groupe de Recherche et d'Echanges Technologiques, 203 Rue Lafayette, Paris 75010, France.

IT
Intermediate Technology, Myson House, Railway Terrace, Rugby CV21 3HT, UK.

NRI
Natural Resources Institute, Central Avenue, Chatham Maritime, Chatham, Kent ME4 4TB, UK. (Formerly TDRI)

FAO
Food and Agriculture Organization of the UN, Agricultural Services Division, Via della Terme di Caracalla, 00100 Rome, Italy.

Central and South America

CEMAT
Centro Mesoamericano de Estudios Sobre Tecnología Apropriada, Apartado Postal 1160, 18 Calle 22–25 Zona 10, Ciudad Guatemala, Guatemala.

INCAP
Instituto de Nutrición de Centro América, Apartado Postal 1188, Carretera Roosevelt Zona 11, Ciudad de Guatemala, Guatemala.

CITA
Centro de Investigaciones en Tecnologías de Alimentos, Universidad de Costa Rica, San José, Costa Rica.

CIAT
Centro Internacional de Agricultura Tropica, P.O. Box 6713, Cali, Colombia.

ITINTEC
Instituto de Investigacion Tecnología, P.O. Box 145, Lima, Peru.

Pacific

ATDI
Appropriate Technology Development Institute, The Papua New Guinea University of Technology, PO Box 793, Lae, Papua New Guinea.

ATC
Appropriate Technology Centre, College of Agriculture Complex, Xavier University, Cargayandeoro City, The Philippines 8401.

North America

Post Harvest Institute for Perishables Information Centre
314 University of Idaho Library, Moscow, Idaho 83843, USA.

Brace Research Institute
McGill University, Montreal 2 PQ, Canada.

IDRC
International Development Research Centre, Box 8500, Ottawa, Canada K1G 3H9.

UNICEF
United Nations Fund for Children, 886 United Nations, New York, NY 10017, USA.

NMSEA
New Mexico Solar Energy Association, P.O. Box 2004, Sante Fe, NM 87501, USA.

VITA
Volunteers in Technical Assistance, 1815 North Lynn Street, Arlington, VA 22209, USA.

ATI
Appropriate Technology Institute, 1331 H Street, NW, Suite 1200, Washington, DC 20005, USA.

Pueblo a Pueblo
1616 Montrose Blvd., Houston, TX 77006, USA.